Critical Perspectives on Social Constructions of Nature

This book explores the commodification of nature with a particular emphasis on how this process is perpetrated by capitalist societies.

Through the example of Italy, the book presents another lens to observe landscapes and natures and the different actors, both human and non-human, involved in their production. It draws on critical geography perspectives to unveil the socio-natural dynamics that form the basis of the commodification of Italian natures. It examines how the social construction of nature in Italy produces peculiar effects on ecologies and landscapes as well as on human and non-human bodies, and how alternative socionatures emerge.

The book is written primarily for scholars, particularly in human, cultural and critical geography, political ecology, cultural studies as well as heritage, tourism, sustainability and urban studies. It will also be of interest to policymakers and planners in nature conservation, landscape planning and heritage management.

Sara Bonati, PhD in Historical, Geographical, Anthropological Studies (curriculum in Physical and Human Geography), is Assistant Professor in Human Geography at the University of Genoa (Department of Education – DiSFor) where she teaches Social Geography, Geography and Communication, and Tourism Geography. She is also associate researcher at the CIERL – Centro de Investigação em Estudos Regionais e Locais (University of Madeira). Her main research areas are disaster studies, urban studies, climate change adaptation, and vulnerability studies, especially in relation to contexts exposed to tourism pressure and landscape transformations for tourism. She participates in the informal group geo.naturae, which brings together Italian researchers who work on socionatures in Italy. She has participated in national and international research projects with different roles (e.g., LINKS H2020 European Project). She is the author of several papers in national and international journals and she has been co-editor of the Routledge book *Governance of Risk, Hazards and Disasters: Trends in Theory and Practice* (Forino G., Bonati S., Calandra L. 2018).

Margherita Cisani, PhD in Historical, Geographical, Anthropological Studies (curriculum in Physical and Human Geography), is Assistant Professor of Economic-Political Geography at the University of Padua (Department of Historic and Geographical Science and the Ancient World – DiSSGeA) where she teaches

Tourism Geography Heritage and Sustainability. Her research focuses on the interplay between landscapes and mobilities in everyday contexts, in natural and cultural heritage landscapes as well as in formal and non-formal educational practices. She has been P.I. of the research project "Landscape Education, Heritage and Justice" funded by the Landscape Research Group in 2019. She is a member of the "Landscape Studies" group of the Italian Association of Geographers and of the " MOHU – Mobilities and the Humanities Centre for Advanced Studies" of the University of Padua where she acted as convenor between 2019 and 2021. Among her publications: Cisani M., Castiglioni B., Di Matteo G. (2025), *Patrimoni oltre natura e cultura. Le riserve della Biosfera Unesco in Italia*, Mimesis.

Marco Tononi, PhD in Human and Physical Geography, is Assistant Professor of Economic-Political Geography at the Department of Foreign Languages, Literatures and Cultures of the University of Bergamo, where he teaches Geography of Movement, Environmental and Landscape Protection, Territorial Governance and Cultural Geography. His research interests concern urban geography, urban natures, urban political ecology, Geographic Information System and participatory cartography tools. He is a member of the "Environmental Geography" group within the Italian Association of Geographers. He has participated, covering different roles in some research projects, at regional and national levels, concerning cultural sustainability, participatory mapping processes, cultural ecosystem services, and nature-based solutions. He is the author of several papers in national and international journals. Among his publications: Tononi M. and Pietta A. (2020), *Città in transizione: un'analisi geografica delle relazioni socio-ecologiche a scala urbana*, Mimesis.

Giacomo Zanolin, PhD in Cultural Heritage and Environment at the University of Milan, is currently an Associate Professor in Human Geography at the University of Genoa, Department of Educational Sciences. At the University of Genoa he teaches: Geography; Cultural heritage and tourism geography; Communication geographies and touristic publishing; Didactics of geography; and Geography of Italy and Europe. He is deputy-coordinator of the master degree in Development of Territories and Sustainable Tourisms at the University of Genoa. He is Member of the Steering Committee of the IGU Commission on the Sustainability of the Rural Systems (IGU-CSRS), President of the regional section (Liguria) of the Italian Association of Geography Teachers (AIIG) and coordinator of the national working group "Geography and Literature" of the Italian Association of Geographers (AGeI). His research interests are mainly focused on: tourism in protected areas and rural spaces; geo-literary narratives and representations; and didactics of geography. In 2023 he won, as head of unit, a Research Project of National Interest (PRIN) titled "Re-Visualising 'the West': geo-literary images of Europe in contemporary Sinophone writings". Among his publications: Zanolin G. (2022), *Geografia dei Parchi Nazionali Italiani*, Carocci.

Critical Perspectives on Social Constructions of Nature

Italy and the Bel Paese

Edited by Sara Bonati, Margherita Cisani, Marco Tononi, and Giacomo Zanolin

Routledge
Taylor & Francis Group
LONDON AND NEW YORK

First published 2025
by Routledge
4 Park Square, Milton Park, Abingdon, Oxon OX14 4RN

and by Routledge
605 Third Avenue, New York, NY 10158

Routledge is an imprint of the Taylor & Francis Group, an informa business

British Library Cataloguing-in-Publication Data
A catalogue record for this book is available from the British Library

ISBN: 978-1-032-73626-6 (hbk)
ISBN: 978-1-032-73631-0 (pbk)
ISBN: 978-1-003-46516-4 (ebk)

DOI: 10.4324/9781003465164

Typeset in Times New Roman
by SPi Technologies India Pvt Ltd (Straive)

Contents

List of figures *viii*
List of tables *ix*
List of contributors *x*

Introduction: Geographies and natures: Facing the social construction of nature in Italy 1
SARA BONATI, MARGHERITA CISANI, MARCO TONONI, AND GIACOMO ZANOLIN

PART I
Heritagisation of nature: Building the Italian brand of nature **13**

1 National parks, nationalism, and the construction of the idea of nature in Italy: Reflections on the Abruzzo, Lazio, and Molise National Park 15
GIACOMO ZANOLIN

2 Transcending nature conservation in Italian UNESCO biosphere reserves 26
MARGHERITA CISANI

3 Rewilding natures in the Ligurian mountains 37
PIETRO PIANA, LORENZO BROCADA, REBEKKA DOSSCHE, AND ENRICO PRIARONE

4 Nature in food and food in nature: Representing, inventing and practising naturalness and mountainness 49
GIACOMO PETTENATI

5 Authenticity (re-)naturalised: The mise-en-scene of raw meat in the heritagized city of Florence 61
PANOS BOURLESSAS AND MATTEO PUTTILLI

PART II
Narratives of nature in the contemporary environmental crisis **71**

6 The vanishing shapes of Dolomitic ice: Temporalities and aesthetics of Marmolada glacier on the threshold of its extinction 73
MAURO VAROTTO AND MATTEO VIANELLO

7 The commodification of mountain natures in the eco-climate crisis: The Milano Cortina 2026 Olympics and Dolomites 82
ANDREA ZINZANI

8 The co(a)sts of beauty: From the construction of the tourist imaginary of attractiveness to increased exposure to risk 95
SARA BONATI, ELEONORA GIOIA, AND ELEONORA GUADAGNO

9 The expansion of agrivoltaics in Italy: A sustainable panacea or a new form of green grabbing? 108
FRANCESCA PERONI, ANNACHIARA ROSATI, VIRGINIA VONA, AND DANIELE CODATO

PART III
More-than (peri)urban natures **123**

10 Urban socionatures and climate strategies: The cases of three medium-sized cities in Northern Italy 125
MARCO TONONI

11 Greening the "void": Brownfield redevelopment and soil desealing in two Italian cities 139
EMANUELE GARDA

12 (Peri)urban and regional protected areas: Structure and prospects of the sociopolitical nature of conservation in Lombardy 151
MARGHERITA CISANI, MARCO TONONI, AND RENATO FERLINGHETTI

13 Wild encounters: Toward a reinterpretation of urban green policies by the rivers in Rome and Cagliari 163
MARTINA LOI AND GINEVRA PIERUCCI

14 Performing animality in city branding: The case of the *Sartiglia* in Oristano (Sardinia) 174
GIAMPIETRO MAZZA AND SARA BONATI

PART IV
Geo-literary representations of Italian natures **189**

15 Il Bel Paese by Antonio Stoppani: A geoliterary approach 191
DAVIDE PAPOTTI

16 "*Mare crespo, vento fresco*": Evaluation of the seascape in *I Malavoglia* by Giovanni Verga 200
ENRICO SQUARCINA

17 The Italian countryside between rurality and nature: Wandering around Cesare Pavese's hills and novels 212
DINO GAVINELLI AND GIACOMO ZANOLIN

Conclusion: Italian socionatures 223
FILIPPO MENGA

Index *230*

Figures

2.1 Map of the Italian BR boundaries and zoning. Cartographic elaboration by Francesco Ferrarese on data collected by the author 30
3.1 The evolution of woodland cover in Liguria between 1936 (Italian Kingdom Forest Map), 1990 (CORINE Land Cover) and 2018 (Ligurian Land Use Map). Made by the authors with QGIS. Sources: Ferretti et al., 2018; Copernicus, 1990; Regione Liguria, 2018; ESRI 40
5.1 A tourist taking pictures of the raw meat exposed in the windows of one of the most iconic *bistecca* restaurants of the UNESCO area. Authors' picture 62
7.1 Olympics poster in Cortina highlighting the prevailing discourse. Source: author, 2022 87
8.1 Map of Senigallia: (i) inundated areas after the flooding of the river Misa on 15–16 September 2022, (ii) land use, and (iii) shoreline dynamics between 2006 and 2020 (elaboration by the author, based on the Copernicus Emergency Management Service—Mapping and ISPRA) 99
8.2 Coastal area of Senigallia in 2023 (Google Earth); the shaded polygon indicates the extent of the inhabited centre in 1889, based on the authors' elaboration of data from Santoni & Morici (2021) 100
8.3 Barano d'Ischia (NA), Maronti beach. Debris fall. Photo by the authors (2023) 101
9.1 Chart illustration of the number of articles by year, country, and subject area 111
9.2 Distribution of the number of project developers (legal address) and EIA projects per Province 117
10.1 Examples of new and old urban socionatures in the case studies. Photographs by the author 135
12.1 Degrees of urbanisation near and inside Lombardy's protected areas 155
14.1 The itineraries of the *Sartiglia*. Authors' elaboration with google earth pro 183
14.2 Image of the Sartiglia. Source: Archivio Fondazione Oristano 184

Tables

2.1 Italian BRs. elaboration by the author on data from UNESCO, MASE and official BR websites 31

3.1 Comparative diagram of agricultural land use categories in the 1929 Agricultural Land Register and Agricultural Censuses of 1990 and 2020 41

9.1 Summary table of Italian laws that refer to agrivoltaics 113

10.1 The three climate strategies analysed: partners, funding and main actions 130

Contributors

Sara Bonati, PhD in in Physical and Human Geography, is Assistant Professor in Human Geography at the University of Genoa (Department of Education – DiSFor). She is also associate researcher at the CIERL – Centro de Investigação em Estudos Regionais e Locais (University of Madeira). Her main fields of research are social vulnerability, disaster risk reduction, and climate change adaptation.

Panos Bourlessas is a human geographer, lecturer at the Department of History, Archaeology, Geography, Fine and Performing Arts of the University of Florence, where he teaches Urban Geography and Social and Cultural Geography. His interests revolve, among others, around geographies of care, studentification, and foodification.

Lorenzo Brocada is research fellow at University of Genoa and PhD student in Cultures, Literatures, Tourism and Territory at the University of Sassari. He did a visiting at the Laboratoire ESPACE of the University of Aix-Marseille (France). His research interests concern environmental and landscape issues from a geographical perspective with a focus on the Western Mediterranean region.

Margherita Cisani, PhD in Historical, Geographical, Anthropological Studies, is Assistant Professor of Economic-Political Geography at the University of Padua (Department of Historic and Geographical Science and the Ancient World). Her research focuses on the interplay between landscapes and mobilities in everyday contexts, in natural and cultural heritage landscapes and in educational practices.

Daniele Codato (PhD in Geography) is a postdoc researcher at the University of Padova. His experience in GIS, PGIS, and remote sensing tools is aimed at investigating socio-cultural and biological diversity and ecosystem services in urban and remote areas.

Rebekka Dossche is an assistant professor in Geography at the University of Genoa (DAFIST). She has a PhD in History at the University of Genoa (Italy) and in Science at the University of Ghent (Belgium). Her research topics concern quanti- as qualitative analysis, historical geography, rural development, ecosystem and landscape services in mountain and remote areas.

Renato Ferlinghetti, geographer and naturalist, is a professor at the University of Bergamo. His research focuses on landscape, urban regeneration processes, environmental management and protection. For ten years councillor of the Colli di Bergamo Regional Park, coordinator of the cultural path towards the signing of the Bergamo-Brescia Charter of protected peri-urban areas.

Emanuele Garda is a Town planner and Associate Professor in Urban and Regional Planning at the University of Bergamo. He holds a PhD in Urban and Regional Planning at the Polytechnic of Milan. His research deals with urban regeneration process, soil renaturation interventions, and correlations with urban planning tools.

Dino Gavinelli is a full professor in Human Geography at the University of Milan where he teaches Regional and Human Geography and Cultural Geography. He was a visiting professor in the Universities of Genoa, Venice, Paris VII, Paris X and Artois (France), Dalian (China). The main fields of his research regard regional studies, urban geography, cultural geography, geography of the rural spaces, cartography, and semiology.

Eleonora Gioia is a Research Fellow in Geography at the Department of Life and Environmental Sciences, Polytechnic University of Marche. Holding a PhD in Civil and Environmental Protection, her research focuses on disaster risk reduction, climate change adaptation, vulnerability assessment, and community resilience. She has worked on several projects addressing resilience and environmental challenges.

Eleonora Guadagno (PhD in Geography from the University of Poitiers, France) is currently a researcher at the University of Naples "L'Orientale", where she teaches Human Geography and Geography of Asia and Africa. Her research interests include, among others, environmental geography, with a particular focus on agricultural and coastal areas.

Martina Loi is a post-doc researcher at the Department of Political and Social Science at the University of Cagliari. She is interested in the relation between urban infrastructures and local urbanisation processes, and in the analysis of urban policies through an urban political ecology lens.

Giampietro Mazza is Assistant Professor in Geography at the Department of Education Sciences of the University of Genoa. His research interests focus on the analysis of socio-cultural and identity dynamics in marginal areas, sport and power geographies, especially in relation to carceral and more-than-human geography.

Filippo Menga is Associate Professor of Geography at the University of Bergamo and Editor-in-Chief of the journal Political Geography. He worked at several European institutions, including the University of Reading and the University of Manchester. He has authored numerous articles published in leading international journals in the fields of geography and political ecology. He is also the author of Thirst (Verso Books), Power and Water in Central Asia (Routledge).

Davide Papotti, after completing a M.A. in Italian Studies at the University of Virginia and a PhD in Geography at the University of Padua, is currently professor of Geography at the University of Parma, Italy. His main research interests are the relationships between geography and literature, place and tourism marketing, the cultural geography of the river Po area (in Northern Italy).

Francesca Peroni (PhD in Geography) is a MSCA postdoc researcher at the University of Padova. She works on urban sustainability, with a focus on soil sealing by adopting a quanti-qualitative approach in the framework of GIScience. Recent research activities are related to climate justice in urban context aimed at investigating how climate change can create spatially differentiated and unequal impacts on the urban terrain and communities.

Pietro Piana is assistant professor in Political and Economic Geography at the University of Genova (Department of Political and International Sciences). He holds a PhD in Geography at the University of Nottingham (UK). His research interests concern the study of landscape change through time in rural and urban areas and its current physical, social, and economic effects.

Ginevra Pierucci is a PhD candidate in Social Geography at the University of Padua and Cà Foscari in Venice and member of the International Project of Mobility and Humanities. Her PhD research focuses on Urban Wildness through the lens of Mobilities Studies, working in the field by the embodied practice of crossing and a more-than-representational approach.

Enrico Priarone is PhD candidate in Security, Risk and Vulnerability at the University of Genoa, where he formerly was research fellow. His current research in rural geography focuses on inner areas and the transformation of landscape practices, with a particular emphasis on organic farming. He also published papers on rewilding, administrative geography and Mega Events.

Matteo Puttilli is Associate Professor in the Department of History, Archaeology, Geography, Fine and Performing Arts at the University of Florence, where he teaches Historical and Political Geography. His research interests involve the analysis of social transformations in cities and geography education.

Giacomo Pettenati is Assistant Professor in Geography at the Department of Economics and Business Studies of the University of Eastern Piedmont, in Novara. He holds a PhD in Environment and Territory at the Polytechnic University of Turin. His main research interests are the geographies of food, spatial processes in rural and mountain regions, landscapes heritage-making.

Annachiara Rosati is a PhD student at the University of Bari, specialising in the study of deep time and the promotion of geoconservation for the sustainable development of protected areas of geological significance. Her research focuses on advancing geological outreach while also addressing geosciences ethics and history.

Enrico Squarcina is full professor of geography at the University of Milan Bicocca where he teaches Cultural Geography and Geography and Didactics of Geography. His fields of research include human geography of the sea and the didactics of geography.

Marco Tononi, PhD in Human and Physical Geography, is Assistant Professor of Economic-Political Geography at the Department of Foreign Languages, Literatures and Cultures of the University of Bergamo. His research interests concern urban geography, urban natures, urban political ecology, Geographic Information System and participatory cartography tools.

Mauro Varotto is full professor of Geography and Cultural Geography at the University of Padova. His research topics concern the evolution of the contemporary Italian mountains. As a glaciological operator he has been carrying out measurements on the Marmolada glacier since 2005.

Matteo Vianello is a Research fellow at University IUAV in Venice and adjunct professor at the University of Genova, Department of Architecture. His field of study are the narratives and representation of design and landscape in relation to climate crises.

Virginia Vona works as an analyst for an environmental consulting firm, specializing in assessing corporate impacts on biodiversity and climate. She focuses on impact evaluations and carbon footprint calculations, using GIS tools to analyze spatial data and support environmental projects.

Giacomo Zanolin, PhD in Cultural Heritage and Environment at the University of Milan, is Associate Professor in Human Geography at the University of Genoa, Department of Educational Sciences. He is also Member of the Steering Committee of the IGU Commission on the Sustainability of the Rural Systems (IGU-CSRS).

Andrea Zinzani is Assistant Professor at the the Department of History and Cultures, University of Bologna. PhD in Human Geography at the Universities of Fribourg (CH) and Verona, his research contributes to geographies of the environment and political ecology and has focused on politics of water resources in Central Asia, coastal governance in South-east Asia and on heritage-scapes and mountain environmental change.

Introduction

Geographies and natures: Facing the social construction of nature in Italy

Sara Bonati, Margherita Cisani, Marco Tononi, and Giacomo Zanolin

Volume premises

This volume has its origins in long-standing dialogues between academics involved in the debate on social nature in Italy. This is a continuation of the first collective work proposed in the Italian language in 2021 (Bonati et al., 2021) to revitalise the discussion on the production of nature, focussing on how capitalist societies perpetrate this process, and it is specifically grounded in Italy. With our first joint special issue, we maintained that in Italy the phenomenon produces peculiar effects on ecologies, landscapes and human and non-human bodies, from which peculiar socionatures emerge, as a result, according to Swyngedouw (1999, p. 446), of a "historical-geographical process" that "insists on the inseparability of society and nature and maintains the unity of socionature as a process". Thus, thanks to a further collective effort, the results of which are far richer than we could have expected, this work aims to present to an international audience and from a critical perspective a series of cases of transformation and commodification of nature in Italy, also taking into account the intensification of global environmental challenges and their effects on the country.

The discussion on the social role of nature has deep origins in academic debate. An important boost to the debate was provided in the nineteenth century in Jena during the meetings between Friedrich Schiller, Johann Wolfgang von Goethe and the von Humboldt brothers, Wilhelm and Alexander. Already on those occasions, these illustrious intellectuals (think about Schelling's Naturphilosophie), discussed the meaning to be attributed to nature (Wulf, 2015). The inspiration was provided by the publication of the first two of Kant's *Critiques*, in which Kant poses the question of whether the order of the world—including human action—can be traced back to universal and necessary laws, thus ultimately to an order inherent in nature itself, or whether mankind is free and thus capable of constructing the meaning of things in its own mind through its own moral actions. But it is in his third critique, the *Critique of Judgement*, that Kant attempts to reconcile the two spheres: that of nature, governed by laws, and that of human freedom, engaged in moral activity through the acknowledgement of the feeling that makes us discover a specific

DOI: 10.4324/9781003465164-1

aim in nature. The topic is thus the relationship between human reality and (we would say today) the social construction of nature. It is, then, no coincidence that in the monumental work that Alexander von Humboldt decided to call *Cosmos*, he argues for a substantial continuity between the shadowy feeling of the intimate bond that binds all things together (and that we feel in our minds) and the order of the cosmos (Serra & Tanca, 2023; Tanca, 2012); between the one and the other there is a structural link, which it is the task of science to clarify (Humboldt, 1845/1997).

The discussion continued in several ways during the nineteenth century, finding specific arguments in the contexts of Marxist and anarchist social struggles. A relevant contribution was made by anarchist geographer Élisée Reclus, who strongly refused the separation of society and nature, according to Spinoza's school of thought, and questioned the supposed superiority of the first over the second. Furthermore, Reclus developed a consubstantiality theory between humanity and nature that anticipated most hybrid and more-than-human contemporary thought. In this perspective, Reclus considered the concept of wilderness as a constructed means of separation and asked to work for a collaborative relationship between society and what was at that time called nature (Ferretti, 2022).

On the other hand, in the context of reflections on Marxist thought, about a century later, in 1984, Smith offered a rereading of Marx about the relationship between capital and nature. In *Uneven Development: Nature, Capital and the Production of Space*, Smith (1984) highlights how Marx outlined two dimensions belonging to nature, "the ideology of nature" that is connected to its representation according to the bourgeoisie and "the destruction of nature" perpetrated by capitalism that is the basis of the conceptualisation of two rhetorics that are still alive in contemporary society: "saving nature(s)" and "managing nature(s)". Similarly, in *The Iconography of Landscape,* Denis Cosgrove and Stephen Daniels (1985) adopt a Marxist interpretation of landscape as a specific way to see and represent the relationship between nature and society imposed by the bourgeois class (Wylie, 2007).

In the 1990s, the debate was intensified with a focus on how nature is (re)produced and on the power dynamics related to who controls this process (Whatmore & Boucher, 1993). In geography, at the turn of the new millennium, following Raymond Williams' thought (Giblett, 2012), Castree and Braun fed one of the most relevant debates that has strongly animated this discussion. They published several works, including edited volumes (i.e., Braun & Castree, 1998; Castree & Braun, 2001) and monographs (Castree, 2005, 2014), highlighting how the debate is still alive and needs to be explored to unravel what they call "epistemic workers" (all those actors who produce and disseminate specific perspectives on things and processes that condition the way we look to them) and "processes of naturalisation" (how the concept of nature is used to reduce the disputes on specific ideologies/interpretations of reality and to deresponsibilise the society). Further works were then produced, opening new perspectives, like that offered by hybrid and more-than-human geography, that have been fundamental to reducing the anthropocentric perspective coming from naturalisation (e.g., Whatmore, 2002).

These premises were not intended to be exhaustive of a long and complex debate (think, e.g., of the contributions provided by David Harvey, e.g. 1996, Jason Moore, e.g. 2016, but also by the political ecology school and its urban group, with particular attention to Erik Swyngedouw's work) but to highlight, in some of its fundamental stages, the evolution of the debate and the theoretical bases on which this volume is based.

Thus, more than a century and a half after the publication of *Cosmos*, academic geography has explored the farthest corners of the planet and developed the most disparate methodologies, splitting into an infinite number of branches. Nevertheless, the original question remains the same, and the need for an up-to-date answer has not gone away as new variables and perspectives have fed the debate. In the meantime, the world has changed, and we have entered what has been termed the "Capitalocene" (Moore, 2016) as the impacts of human action, specifically capitalism, on ecosystems, amplified by technical and technological developments over the last two hundred years, have become increasingly evident. What George Perkins Marsh (1864/2003) had begun to observe at the end of the nineteenth century (among other things, based on a journey that had a very important stop in Italy), especially that humans are part of nature and have an unbalancing power on it, is today an evidence. However, a lot of work is still required to address this issue.

Accordingly, this book explores how Italian natures are shaped, constructed and commodified through policies, narratives and practices. Thus, we shall begin with the call for denaturalisation as set forth by Demeritt (2002, p. 769), who states that "one objective of denaturalisation is to show that something is bad and that we would be better off it were radically changed, which becomes conceivable once we realise it is socially constructed and within our power to change". Thus, the central question addressed in this book is how the construction of nature has contributed to and continues to influence the commodification of nature in Italy. This requires examining the mystifying strategies employed to perpetuate this process and their associated consequences.

We acknowledge that in doing this we limit our discussion to a specific geographical border, an overall context of analysis that characterises all the chapters: Italy. However, at the same time, we are aware (and this is one of our purposes) that these geographical borders are permeable and fictitious. As Armiero and Hall (2010, p. 2) state when talking about Italian environmental history, "Italian environmental history has never been just Italian"; similarly, we could say that Italian nature has never been just Italian in terms of ecological, social and disciplinary boundaries. Thus, with this statement, we recognise that several international actors have had a central role in feeding these processes of the commodification of Italy, and especially of its beauties (think, e.g., to the role of the UNESCO brand in the celebration of the landscape of the "Prosecco Hills" of Conegliano and Valdobbiadene), producing a naturalising narrative that ultimately justifies their capitalisation.

Furthermore, in this book, we examine how Italian natures can be considered paradigmatic in the analysis of the global natures in the Capitalocene, starting from the recognition that different national and international discourses and actors continuously appropriate the idea of Italy, of its nature and beauty, and, at the same time, they have highly contributed to shaping it.

The beautification of Italy

The idea of "Bel Paese" (meaning "land of beauty") and its consistent use in association with Italy is emblematic of what is stated at the end of the previous paragraph. Even though this expression has literary origins in the thirteenth and fourteenth centuries, particularly in Dante Alighieri's *Divina Commedia* and Francesco Petrarca's *Canzoniere*, its contemporary use is directly linked to the construction of so-called "made in Italy" as a global brand started in the second half of the twentieth century.

During this last period, the expression "made in…" took on a particular meaning throughout the world, establishing itself not only as a certificate of the origin of a product but also as a tool to represent the image of the country of production. On this, marketing studies have largely investigated the so-called "country of origin effect" (Jaffe & Nebenzahl, 2006), which refers to the ability of a label to generate commercial added value derived from the association between the product and the place of production, based on a collective imagination that associates certain forms of production excellence with specific countries.

The commercial strategies linked to "made in Italy" originated in the 1970s. During that period, the crisis of large-scale industry due to the global oil crisis opened up space for the growth and affirmation of industrial districts located in specific Italian regions based on artisanal skills rooted in local history, which after the Second World War had evolved into small-and medium-sized Italian enterprises (Becattini, 1998, 2000). The result, especially in Northern Italy, was the development of a very particular form of capitalism, which has been defined as "molecular" (Bonomi, 1997), to describe its density and capillarity.

Over the decades, "made in Italy" has become a tool for building added value for Italian goods based on a principle, defined arbitrarily, according to which Italy is the homeland of beauty and good taste (Belfanti, 2019). This association between "Italianness" and a predisposition for beauty and refinement in taste has been consolidated over the last few decades, to the point of becoming an axiom that is difficult to question. "Made in Italy", therefore, no longer describes only a concrete value but also a symbolic one. In this process, Italy itself has become a product to sell, and this effect is no longer limited to its goods, having also pervaded landscapes and communities that have become part of the brand.

Nowadays, the sense of beauty is still considered typical of the Italian style, which is inspired by the country's landscapes and natural beauties. Thus, the Bel Paese can be considered one of the main brand campaigns that Italy has promoted and that continues to receive consent at the national level and abroad. Italian national TV channels, for example, systematically promote shows that spread and talk about the beauty of Italy and its natural heritage (alongside its renowned cultural heritage). Even from an institutional point of view, nature conservation has its origins primarily in its recognition of natural beauty, and this bias is still present today. This is part of a continuous process of beautification of the country that has the purpose of both attracting interest in it and shaping reality, making it adherent to the imaginary according to place branding strategies (e.g., Colomb & Kalandides, 2010).

This has produced the activation of continuous interventions, realised as part of the beautification process, used to change the aspects of places, making them more attractive to both internal and external observers and introducing elements that contribute to improving their appearance. However, these processes, as better described to follow, can produce physical changes from which there is no turning back and that, thus, could have high costs for those undergoing this "treatment".

To better understand this, an example coming from the cinema could help. In the Oscar-winning movie *The Great Beauty* (2013), directed by Paolo Sorrentino, the beautiful landscapes and architecture of Rome are the stage on which the different characters interact. Beauty in the film is an inherent characteristic of Italian society and also of Italian landscapes, a way of life for the Italian people. Behind the beauty, however, there is a series of contradictions that emerges during the film: the beauty of the landscapes, in particular, contrasts with certain socio-cultural processes of decay within society, such as the contrast of architectural beauty and the inability to enhance it, or the contradictory nature of the contemporary social practices of mass tourism and social evenings. What therefore emerges from the film is how the ostentation of beauty and its idea can hide the decadence of the system on which it feeds.

This can be better presented by borrowing language from aesthetic medicine, and in particular the idea of the "botulinisation of bodies", meant in that context as the process of applying botulinum injections to obtain an aesthetic effect through muscle relaxation. Indeed, it describes an action intended to improve appearance by applying a limited, well-defined quantity of botulinum neurotoxin. Botulinum toxin has been widely promoted as a beauty elixir, but it can also be perilous, even deadly, if taken in the wrong way. As recommended in aesthetic medicine, these treatments should be applied only by expert medical personnel, and only after a careful medical history is taken, to be considered safe. Furthermore, this allows one to choose the most suitable treatment based on the type of blemish being treated and the patient's own expectations.

This image taken from aesthetic medicine is useful for understanding the process of decadence, introduced above, and its potential, and in some cases irreparable, consequences on landscapes and places in Italy; a sort of "botulinisation of places". With this, we refer to all those processes (equivalent to botulinum injections) that incorporate and activate a whole host of different branding practices and strategies to maintain the appearance of great beauty, even when it has disappeared or been significantly compromised, that could degenerate into situations of obsession and/or perversion. For instance, some urban regeneration projects create new, smart or eco district inside the Italian urban areas, that, in some cases, can be conceived as external elements in the body of the city, disconnected by the others; sorts of "pillow face syndromes", using newly the language of aesthetic medicine.

Furthermore, as in aesthetic medical practice, for landscapes, it should be required to use these strategies carefully and in a controlled manner. Just as overuse or misuse of botulinum in aesthetic medicine can produce deformation in a body, with damage sometimes irreparable and including, in extreme cases, death of the body, similarly, we can observe and discuss this process of targeted destructive interventions on Italian ecological systems, natural landscapes and heritage. These practices, like botulinum, have the power to eliminate "facial creases" (e.g., "wrinkles", "ugliness"), which socially have been defined as bad, like producing the expulsion of marginalised people or reducing them to the silence, according to the practices of shadowing and ghosting largely described by critical branding theory (Short, 1999; Vanolo, 2017). Furthermore, they can be responsible for the flattening of differences and the homologation of places to a unique standard, banal-ising and stereotyping the landscapes and their values and meanings. Thus, these systems are often far from being fair and just, instead generating poverty, reducing rights and complexity and ultimately producing (interspecific) inequalities. In this idea, there is also the dimension of the anesthetisation and/or annihilation of a place's life, that could degenerate, in the most extreme forms, like gentrification and/or deresidentialisation of places and final loss of their social function. However, at the moment, the cost of these treatments is poorly discussed and problematised in Italy, which could be also considered a way to protect the dominant narrative.

Nature and natural elements are often used as a means of beautification, sometimes becoming examples of place botulinisation. This is at the basis of processes of the commodification and capitalisation of nature, such as overconsumption, according to neoliberal logic, that reiterate a dual approach to society and nature. An example of this process is the creation of parks and green spaces seen as an improvement to the quality of life of the inhabitants and usually considered an element of place branding and territorial marketing. However, behind these projects of renaturalisation, several critical studies unveil the moral intention of this green beautification, to modernise and spread middle-class lifestyles without really solving local social and environmental challenges, potentially even escalating them. The planning and implementation of new parks could generate spatial injustices, such as the displacement of certain classes of inhabitants or the problem of unequal access to these parks (Angelo, 2020; Gandy, 2005, 2022; Zimmer et al., 2017).

The example of nature in the urban context shows how experimental interventions to improve the environmental characteristics of cities and nature-based solutions, often for place branding and marketing reasons, actually produce other effects, such as improving the appearance of some specific and limited parts of a city but also increasing the distance from other parts in terms of spatial inequalities and differences in opportunities and the quality of urban spaces. This eco-technical approach of using natural elements to improve the ecological appearance of a part of the city is not for everyone: it is for a limited, selected population.

Thus, today as in the past, we can say that the idea of Bel Paese is maintained and developed according to a capitalist approach to nature. Furthermore, we maintain that these practices reiterate a specific anthropocentric and dichotomous perspective on nature. This Bel Paese narrative is mobilised in regards to both the more explicit capitalist processes of nature exploitation and the discourses geared towards the conservation and protection of Italy's ecological systems and so-called natural landscapes and heritage, that "selling nature to save it" participates in neoliberal processes through the "accumulation by conservation" mechanisms (Büscher & Fletcher, 2014).

Accordingly, denaturalising the process, following Demeritt's suggestion, means for us and for this volume to unveil how specific mechanisms of beautification and promotion of local "nature" are responsible for activating processes of dispossession and/or overconsumption of places with potential negative consequences on their survival, including not only the survival of landscapes and their human and non-human inhabitants but also of their supposed beauty.

Chapter outline and structure

In light of the outlined considerations, this book collects contributions that identify and discuss processes of nature commodification through beautification, or more broadly as part of the Bel Paese narrative, to denaturalise and unravel them. In particular, it focusses on how Italian nature representation and branding (and especially the branding narratives connected to the idea of Bel Paese), are used in defining and reproducing heritage, landscapes and landscape processes. This includes those processes of nature capitalisation and exploitation that are justified under, for example, the sustainability label, such as strategies for climate change adaptation/mitigation and conservation/protection policies (e.g., heritagisation). All the chapters unveil the controversial and unequal effects of these narratives, policies and practices to highlight how nature is always hybrid and built through interactions between humans and non-humans and their agency capacities, the hierarchical power relations existing between them and the consequences of these relations.

This work proposes a series of examples that go beyond the idea of Bel Paese, offering new perspectives on all the different kinds of socionatures that populate Italian landscapes, in both urban and rural contexts and with a transcalar perspective, opening up towards alternatives to the dichotomous and neoliberal relationship between nature and society. All the chapters deal with these points in different ways, approaching specific themes and examples. Although related to specific Italian case studies, each contribution offers valuable insights into other similar contexts and dynamics, especially in the Global North. This book is organised into four sections.

The first section is entitled *Heritagisation of Nature. Building the Italian Brand of Nature*, and it is dedicated to exploring some of the mechanisms through which Italian natures have been and are continuously constructed and reconstructed in relation to the idea of heritage. Nature thus patrimonialised is then used as a brand for political and economic (especially tourist) purposes, linked to the reproduction of power, the maintenance of privileges or the (in) direct perpetration of the dominant capitalist model of relationships with nature. The production of Italy's natural heritage goes primarily through the establishment of protected areas. In this first part of the book, the topic is analysed from a more historical perspective by Zanolin through an analysis of Italian national parks, which, in their establishment, reflect ambiguous objectives between conservation and nationalist celebration. Similarly, Cisani's contribution, delving into the current debate around the tipping points and impasses of mainstream conservation, aims to present a comparative and critical analysis of the Italian UNESCO Biosphere Reserves, which are globally presented as laboratories for biodiversity conservation and sustainable development. Moving towards less institutionalised processes of nature heritagisation, Piana, Brocada, Dossche and Priarone seek to explore the rewilding dynamics and the socio-economic implications of vegetation growth and the return of wildlife in the Ligurian Alps and Apennines. Highlighting how the social construction of nature as heritage also affects other social fields besides conservation in the strictest sense, both Pettenati and Bourlessas and Puttilli address, in different ways, the socio-ecological implications of the supposed naturalness of Italian food. Pettenati investigates how the relations between agrifood production and nature are represented, mobilised and practised in two typical food products' recognition (Protected Designation of Origin labels and slow food), while Bourlessas and Puttilli discuss how authenticity is produced, staged and reproduced through the *mise-en-scene* of foodstuff, focussing on the exhibition of the supposed natural practices of consumption of the famous *bistecca alla fiorentina* (Florentine steak).

The second section, *Narratives of Nature in the Contemporary Environmental Crisis*, continues the discourse introduced in the first section, considering the variables of risk and climate change. This section focusses on how the environmental crisis has been the result of specific narrative choices about nature and how the relationship between nature and society has been built over time. Accordingly, the section starts with the Alps, discussing both the glaciers' status, presented by Varotto and Vianello (with a particular focus on the Marmolada glacier that in the summer of 2022 was the protagonist of a tragedy in which eleven people died), and the conflicts interesting the heritage area of the Dolomites as presented by Zinzani. Here, the narrative of winter tourism and events (particularly the Winter Olympics games as discussed by Zinzani) are used to present and discuss how the natural vocation of the mountain has been built over the years and continues to be reiterated to survive a dying economic system based on heritage exploitation. Similarly, the third chapter, presented by Bonati, Gioia and Guadagno, using the concepts of coastalisation and sand urbanisation as lenses of analysis, investigates how the Italian coasts are at risk today due to climate change and how a continuous process of soil consumption and overexploitation of resources, especially linked to tourism development, has been at the basis of the construction of risk dynamics and coastal beauty disappearance. The last contribution to this section, by Peroni, Rosati, Vona and Codato, takes up soil consumption, focussing on the existing conflicts and tensions around the issues of the mitigation of climate change and Italian landscape preservation, investigating the phenomenon of green grabbing.

The third section, *More-Than (Peri) Urban Natures*, takes up the topic of ecological transformation, focussing on Italian cities and the production of different natures in contemporary urban processes. Overcoming the dichotomy between city and nature, this section investigates the commodification of nature in urban contexts and the urban processes of renaturalisation that involve human and non-human actors. Natural elements are today fundamental in the redefinition of urban policies as tools for improving city branding and image. Accordingly, the opening contribution by Tononi proposes a critical analysis, based on urban political ecology, of the different socionatures produced during the urban socioecological transition to sustainability and climate resilience, focussing on medium-sized, urban, industrial areas in Northern Italy. In the same vein, Garda proposes the identification, analysis and comparison of certain regeneration processes of large brownfield sites, selected in different Italian cities and characterised by the promotion of partial soil desealing interventions and the implementation of nature-based solutions. A different perspective is offered by Cisani, Tononi and Ferlinghetti, who present the history of nature conservation in the Lombard context, focussing specifically on protected areas, such as green belts, around major cities, and the social construction of nature in these processes of the suburbanisation of nature. Similarly, Loi and Pierucci investigate urban green policies and the processes of regeneration of wild and hybrid landscapes in two regeneration projects to create a system of urban parks along the course of the Tiber River in Rome and the Is Cungiaus in Quartucciu (a small town near Cagliari) using a visual and embodied research approach. A different perspective is adopted by Mazza and Bonati, who focus on the role of non-humans in transforming urban spaces and feeding city branding practices. This chapter dialogues with animal and urban geography and, by analysing the power relations between humans and non-humans, aims to highlight the role of the latter in building urban space.

The fourth and last section, *Geo-Literary Representations of Italian Natures*, moves in a direction a bit different in comparison to the previous sections, considering how nature has been narrated and conceptualised in Italian literature and how this has contributed to defining/redefining through time the relationship between society and nature. It explores how this has often generated stereotypes, such as the very idea of the Bel Paese, but also helped to question the dichotomous approach between them. The geo-literary approach at the basis of this section can be considered a relevant method to study how narratives are built and penetrate inside culture, contributing to a shared collective image that can shape reality. Accordingly, the purpose of this section is to go beyond the presumed objective meaning of nature and to explore its socially constructed meaning. More specifically, this section opens with a chapter by Papotti dedicated to the literary construction of the idea of Bel Paese in the foundational book by Antonio Stoppani, which can be read as a document about the conception of nature in the second half of the nineteenth century within the Italian socio-cultural context. The reading of this book is useful to investigate the role that natural heritage has played in the construction of the Italian national identity. Subsequently, Squarcina concentrates on the Mediterranean Sea using a novel by Giovanni Verga (the most important author in Italian verism) to highlight that the sea is an elsewhere that cannot be described simply as a national landscape. Indeed, the chapter discusses its continuously changing meanings. The third and closing chapter of this section, by Gavinelli and Zanolin, presents a critical reading of the novels of Cesare Pavese (a fundamental author in contemporary Italian literature) to reflect on the paradox of the Italian rural landscapes represented at the same time as a legacy of a past based on human labour and as pristine nature.

In conclusion, although this volume cannot be considered thematically or geographically exhaustive in presenting all the ways in which nature is and has been commodified in Italy, it aims to provide a sufficiently broad overview of the different ways in which nature and natural elements continue to be used and consumed according to capitalist logic and how most of these practices continue to feed and/or respond to a Bel Paese project in some way. Thus, the exercise of denaturalisation is considered useful here to unravel the different strategies through which the processes of nature consumption, in association with beautification mechanisms, are naturalised.

References

Angelo, H. (2020). *How green became good, urbanized nature and the making of cities and citizens*. University of Chicago Press.

Armiero, M., & Hall, M. (2010). *Nature and history in modern Italy*. Ohio University Press.

Becattini, G. (1998). *Distretti industriali e made in Italy. Le basi socioculturali del nostro sviluppo economico*. Bollati Boringhieri.

Becattini, G. (2000). *Dal distretto industriale allo sviluppo locale. Svolgimento e difesa di una idea*. Bollati Boringhieri.

Belfanti, C.M. (2019). *Storia culturale del Made in Italy*. Il Mulino.

Bonati, S., Tononi, M., & Zanolin, G. (2021). Social nature geographies: Le geografie e l'approccio sociale alla natura. *Rivista Geografica Italiana*, *CXXVIII*(2), 5–20. https://doi.org/10.3280/rgioa2-2021oa12029

Bonomi, A. (1997). *Il capitalismo molecolare. La società al lavoro nel Nord Italia*. Einaudi.

Braun, B., & Castree, N. (1998). *Remaking reality: Nature at the millenium*. Routledge.

Büscher, B., & Fletcher, R. (2014). Accumulation by conservation. *New Political Economy*, *20*(2), 273–298. https://doi.org/10.1080/13563467.2014.923824

Castree, D. (2005). *Nature*. Routledge.

Castree, D. (2014). *Making sense of nature*. Routledge.

Castree, D., & Braun, B. (2001). *Social nature. Theory, practice, and politics*. Wiley-Blackwell.
Colomb, C., & Kalandides, A. (2010). The 'Be Berlin' campaign: Old wine in new bottles or innovative form of participatory place branding? In G. Ashworth & M. Kavaratzis (Eds.), *Towards effective place brand management: Branding European cities and regions* (pp. 173–190). Edward Elgar.
Cosgrove, D., & Daniels, S. (1985). *The iconography of landscape*. Cambridge University Press.
Demeritt, D. (2002). What is the 'social construction of nature'? A typology and sympathetic critique. *Progress in Human Geography*, *26*(6), 767–790. https://doi.org/10.1191/0309132502ph402oa
Ferretti, F. (2022). Élisée Reclus: A philosophy of nature. *Encyclopédie d'histoire numérique de l'Europe*. https://ehne.fr/en/node/12274
Gandy, M. (2005). Cyborg urbanization: Complexity and monstrosity in the contemporary city. *International Journal of Urban and Regional Research*, *29*(1), 26–49. https://doi.org/10.1111/j.1468-2427.2005.00568.x
Gandy, M. (2022). *Natura Urbana: Ecological constellations in urban space*. The MIT Press.
Giblett, R. (2012). Nature is ordinary too: Raymond Williams as the founder of ecocultural studies. *Cultural Studies*, *26*(6), 922–933.
Harvey, D. (1996). *Justice, nature and the geography of difference*. Blackwell.
Humboldt, von A. (1997). *Cosmos: A Sketch of the Physical Description of the Universe* (vol. 1). Johns Hopkins University Press (original work published 1845).
Jaffe, E.D., & Nebenzahl, I.D. (2006). *National image and competitive advantage. The theory and practice of place branding*. Copenhagen Business School Press.
Marsh, G.P. (2003). *Man and nature: Or physical geography as modified by human action*. University of Washington Press (original work published 1864).
Moore, J. (Ed.). (2016). *Anthropocene or capitalocene?: Nature, history, and the crisis of capitalism*. KAIROS.
Serra, V., & Tanca, M. (2023). Introduzione. In V. Serra & M. Tanca (Eds.), *Raccontare il mondo, descrivere la natura. L'opera di Alexander von Humboldt tra letteratura e scienza* (pp. 7–26). UNICA Press.
Short, J.R. (1999). Urban imagineers: Boosterism and the representation of cities. In A.E.G. Jonas & D. Wilson (Eds.), *The urban growth machine: Critical perspectives two decades later* (pp. 37–54). State University of New York Press.
Smith, N. (1984). *Uneven development: Nature, capital, and the production of space* (3rd ed.). The University of Georgia Press.
Sorrentino, P. (Director). (2013). *La grande Bellezza* [*The great beauty*] [Film]. Indigo Film, Babe Films, Pathé, Medusa Film, France 2 Cinéma. https://www.indigofilm.it/produzione/la-grande-bellezza/
Swyngedouw, E. (1999). Modernity and hybridity: Nature, regeneracionismo, and the production of the Spanish waterscape, 1890–1930. *Annals of the association of American Geographers*, *89*(3), 443–465.
Tanca, M. (2012). *Geografia e filosofia. Materiali di lavoro*. Franco Angeli.
Vanolo, A. (2017). *City branding: The ghostly politics of representation in globalising cities*. Routledge.
Whatmore, S. (2002). *Hybrid geographies: Natures cultures spaces*. Sage.
Whatmore, S., & Boucher, S. (1993). Bargaining with nature: The discourse and practice of 'environmental planning gain'. *Transactions of the Institute of British Geographers*, *17*(2), 166–178. https://doi.org/10.2307/622360
Wulf, A. (2015). *The invention of nature: Alexander von Humboldt's new world*. Murray.
Wylie, J. (2007). *Landscape*. Routledge.
Zimmer, A., Cornea, N., & Véron, R. (2017). Of parks and politics: The production of socio-nature in a Gujarati town. *Local Environment*, *22*(1), 49–66. https://doi.org/10.1080/13549839.2016.1157157

Part I

Heritagisation of nature

Building the Italian brand of nature

1 National parks, nationalism, and the construction of the idea of nature in Italy

Reflections on the Abruzzo, Lazio, and Molise National Park

Giacomo Zanolin

Introduction

This chapter uses an interdisciplinary approach and although it could embrace several disciplines (such as geography, sociology, philosophy, history, political ecology, etc.), it is based on two main theoretical axes, used to develop a coherent way of thinking with regards to the construction of an idea of nature in Italy starting with the example of a national park. First, we take the environmental and cultural branches of geography, whose relevance in the international scientific debate is increasing because of the urgency of topics such as climate change and the relationship between human beings, other living beings, and the biosphere, proposing alternatives to the current socio-ecological crisis based on the convergence between humans and non-humans (Castree & Braun, 2001). The second axis is derived from environmental history, whose relevance in the debate about the social construction of nature is crucial in observing it through the lens of time to study the way in which groups, societies, nations, individuals, and cultures have interacted within the environment in which they live (Armiero & Barca, 2004).

Inside this theoretical framework, this chapter aims to study the complex geographies of Italian national parks and to contribute to the debate on the social construction of nature in Italy, focusing on the ambiguous narratives that precede and follow the creation of the Abruzzo National Park (ANP),[1] one of the first national parks in Italy. Based on the results of an archival research on documents from the first two decades of the 20th century, it includes a discussion of the narratives about Italian national parks and the ways they have contributed to shape the concept of nature in the so-called Bel Paese. In particular, it investigates the ambiguous connection between the ambition to preserve ecosystems and the use of nature as a political tool in the process of forming the Italian national identity (Sievert, 2000; Hardenberg et al., 2019).

The final remarks discuss the relationship between theories and practices in the social construction of an idea of nature in Italy, through the protected areas (PAs), discussing the idea of wild nature and showing that most of the Italian PAs are not pristine but have rural origins. For this reason, the idea that nature, in Italy, is something extra-human that needs to be preserved by excluding humans is wrong, from both practical (Depraz, 2008) and ethical (Berque, 1996) points of view.

DOI: 10.4324/9781003465164-3

Theoretical insights

The geography of Italian PAs is extremely heterogeneous and hard to compress into a concise presentation. However, it is possible to highlight a key regulatory reference: Italian Law 394/1991, which marked a turning point introducing the idea that the PAs are actors that aim to favour the protection of ecosystems and, simultaneously, stimulate the link between social, cultural, political, economic, and ecological dynamics. Within this "integrative" perspective (Depraz, 2008) we find a system composed of 24 national parks, 149 regional and interregional parks, 32 marine PAs and hundreds of regional and state reserves, as well as dozens of wetlands and more than 2500 sites included in the Natura 2000 Network.

As can be surmised, both very small reserves and large national parks are included in this huge system, clarifying the vocation of contemporary Italian PAs as territorial actors, that work not only ecologically but also stimulate sustainable and constructive human activities synergic within other biological actors. Their purpose is therefore not only to preserve ecosystems of unquestionable value, but also to address human societies towards the compatibility between human and non-human needs (Giacomini & Romani, 1982). On these bases, an interpretative approach based on social nature theories is effective to underline the notion that nature is a complex idea (Vidali, 2022), it is neither an objective datum nor a category, but a social product shaped by cultural, political, and economic practices (Castree & Braun, 2001). These reflections are evidently configured within the neo-Marxist theoretical framework and are closely linked to the critique of the social and ecological impacts of capitalism (Smith, 1984/2008).

This chapter is based on these theoretical assumptions and focuses on political aspects related to the role that PAs can play in defining the meaning that a country can attribute to a large part of its territory, as many non-urban regions are often defined, acritically, as natural spaces. This study does not investigate the role of PAs as instruments for planning (although this is obviously a fundamental issue) (Thomas et al., 2003), but their role as actors that generate symbolic meanings that contribute to underpin the political systems of a country (Moschini, 2013). This does not mean underestimating their importance from an ecological point of view (which remains a priority) (Arneth et al., 2023), but emphasising their potentially crucial role as semiotic actors in the broader process of social construction of meanings attributed to territories (Phillips, 2003). The creation of a protected area starts a new phase in the process of territorialisation (Turco, 2010), defining a specific meaning for the concept of nature, which activates the symbolic re-appropriation of space, from which the phases of material and structural control derive. PAs thus define the terms of a contract in which the relationship between humans and other living beings is based on administrative boundaries. Moreover, their ultimate aim is to extend this kind of contract beyond those administrative boundaries, expanding their vision and the resulting policies across the entire territory of a state (Depraz, 2008).

All PAs perform this function, but none as effectively as national parks. This is due to their economic and communicative power, which gives them a greater capacity to populate the collective image of nature, as an effect of the pervasiveness of the North American conception of national parks (Nash, 1967/2014). This is the reason why it is useful to study the historical process of the founding of national parks (Hardenberg et al., 2019; Piccioni, 2023). Understanding the meanings attributed to them at the beginning of their evolution is particularly important for us to see them before the constructed symbolic meaning emerged into an archetypal form, which today we wrongly tend to take for granted. Based on this, the discourses that accompanied the foundation of ANP are particularly interesting, not only because it is the oldest in Italy, but also because it was created at a crucial stage of Italy's history, contributing to the process of national identity construction (Armiero, 2022).

In the case of ANP, as in many others, nature has been narrated since the beginning as pristine (or "virgin") and described as wilderness (Cronon, 1996; Soriani et al., 1996; Tuan, 1974/1990), generating a dichotomic conception of the relationship between humans and non-humans. As we have already noticed (Zanolin & Paül, 2022), the term wilderness does not exist per se, but it depends on the social and spatial context in which it is used; much like nature, wilderness is a social construction (Castree, 2014; Demeritt, 2002). In particular in Italy, the existence of wild spaces, understood as untouched by humans, seems completely meaningless, due to the succession of influx of populations the peninsula has experienced (Agnoletti, 2018). For this reason, it seems more apt to consider so-called natural spaces as nature-society hybrids (Zimmerer, 2000), overcoming implicitly the idea of a condition of stability and natural equilibrium independent of human influence. In Italy, as elsewhere, PAs protect hybrid spaces (Whatmore, 2002) which owe their complexity and rich biodiversity to the centuries-old interaction between humans and non-human ecosystems actors (Büscher & Fletcher, 2020). To hide or show these processes means to address the interaction between local communities and ecosystems in different ways, for this reason it is relevant to study the ideas of nature promoted and symbolised by PAs during the years of Italian national identity formation.

The narratives that led to the foundation of Abruzzo National Park

To study the relationship between state and nation is complex, especially in a context such as Italy where the dynamics that led (in 1861) to the construction of the unified state and the policies through which (in the following decades) the effort to build a national identity had been so hard (Piccioni, 2023; Sievert, 2000). The cultural climate of the time can be summarised through the words of Giovanni Amendola, reported in the magazine "La Voce" on December 1, 1910, on the eve of the celebrations of the fiftieth anniversary of the unification of Italy: "Now, in Italy, this is what we suffer from above all: the nation is little more than a myth that wanes and a hope that rises again".[2] From the 18th century onwards, western Europe developed the idea that the formation of a state needed the construction of a national sentiment on which the sense of cohesion between the members of the community and the recognition of sovereignty needed to be based (Banti, 2011).

Romanticism offered fundamental support to this point of view, including the invention of the figure of the Heimat (the small homeland or the identity territory), which synthesises the feeling of emotional attachment to places of life. The application of this idealised principle to political discourse led, during the 19th century, to the transformation of individual attachment to the place of origin to the collective bond of the members of a nation to the state, presented as the common homeland of its citizens. In this sense, the nation is the embodiment of the spirit of a people, which is expressed using symbolic elements capable of fostering the aggregation of individuals. Several instruments are commonly used by states to build national identities (Thiesse, 1999), and the symbolisation of nature through the Heimat is one of those (Depraz, 2008). The reference to the original characteristics of the nature of a region can contribute to construct a discourse on the place of origin of a nation and thus on the meaning of a people's attachment to its homeland. Nature can be understood, from this point of view, as a symbol, functional to the territorial rooting of the feeling of national identity.

In the 1910s, the parliamentary debate on the need to introduce normative instruments to protect what were defined as "natural beauties" became more and more insistent, as testified by Giovanni Rosadi (1911), who invited parliament to recognise the value of places of "notable public interest because of their natural beauty or their particular relationship with history and literature".[3] The attempt to construct a cultural value for non-urban spaces, through the acknowledgement of their aesthetic value, equating them with artwork, is evident, such as the exaltation of a common culture to instil in the inhabitants of the Italian peninsula the idea of having a common descent: all those are basic elements in the construction of national identity (Banti, 2011). Through its aesthetisation, nature has become a cultural construct that plays a specific role in the process of nation building, such as in many other contexts (Thiesse, 1999). The Italian specificity concerns the increasingly strong correlation built between aestheticised nature and the concept of landscape, understood at the time as an expression of an essentially patrimonial dimension. A proof of this is the law presented by Benedetto Croce in 1920 (and approved in 1922), in which the philosopher and parliamentarian emphasised the need for a regulation that would place "a curb on the unjustified devastation that is being wrought against the most known and loved features of our soil", in order to "defend and enhance, to the greatest extent possible, the greatest beauties of Italy, both natural and artistic", for "high moral reasons and not less important reasons of public economy".[4]

At the origin of the first Italian national parks are therefore discourses based on the aesthetisation and heritagisation of a nature reduced to its landscape dimension, through which it becomes a useful tool in the increasingly urgent process of nation building. Finally, a fundamental aspect should not be overlooked in this process: its promoters were largely urban inhabitants, far removed from the practical thoughts and needs of the inhabitants of the regions identified as "natural" precisely in contrast to urban spaces. The proponents of the Italian national identity therefore had an external and often stereotyped view of the "landscapes" to which they decided to give a role within the heritage of the Italian nation. Obviously, this does not mean that no one was aware of the concrete needs of these regions, but that in the process that led to their construction as symbolic elements (through the foundation of national parks), different figures interacted at various levels, which, more or less consciously (it is difficult to be sure today), assembled the various pieces that made up the specific meaning attributed in Italy to nature as an element of national identification.

The years that prelude the foundation of the national park

A leading role in the process of founding ANP was played by the environmentalist movement, which had begun to develop in the last decades of the 19th century and, in the early years of the 20th century, was increasingly strengthened by some decisive battles, such as that for the defence of the Marmore Falls or the Ravenna Pinewood (Pedrotti, 2012). These were also the years in which the possibility of creating a national park in the upper Sangro valley (nowadays in the administrative region called Abruzzo, central Italy) began to be considered, starting from an area in which a royal hunting reserve had been established (in 1872), and which remained in force until 1912. The exclusive hunting rights of the King in those years favoured the recovery of the population of many species including chamois and bears, but also generated some critical issues on the relationship with the local human population. The tensions led to the closure of the reserve, freeing the hand of hunters, who in a very short time decimated the chamois and bear populations. In this context, a law was passed in 1913 to prohibit the killing or trapping of chamois (Rupicapra ornata). In the same year, the Italian Federation Pro Montibus (IFPM) began to look at this region with the aim of establishing a protected area following the example of the American ones. Among the most engaged personalities were Alessandro Ghigi, Erminio Sipari and Pietro Romualdo Pirotta, who prompted the association to lease, in 1921, 500 hectares in the area of Costa Camosciara, upper Val Fondillo, municipality of Opi (Ferrari, 2015). This was the starting point of a process that lead, in September 1922, to the establishment of a Provisional Directorate (chaired by Erminio Sipari) to manage an area of 12,000 hectares (together with a marginal zone of External Protection of 40,000 hectares), which would be definitively recognised as ANP by Royal Decree on January 11th, 1923.

To understand the reasons for the action of the IFPM, it is possible to consider a key document, the programmatic report "Il Parco Nazionale dell'Abruzzo" by Romualdo Pirotta and published by IFPM in 1917, six years before the establishment of the Park. In this text, the idea that human beings transformed nature for their needs, reducing its beauty, is evident. The shadow of George Perkins Marsh's thinking is also clear: at the end of the 19th century he emphasised the destructive role of human action on ecosystems: "but man has always been and is the first, the greatest, often the fiercest, enemy of nature, of which he is nevertheless a part and of which he claims to be the most distinguished work" (Pirotta, 1917, p. 4). This approach is reinforced by the idea that human action is not only destructive, but also reduces the intrinsic beauty that characterises so-called natural landscapes: "disfigured the landscape, going so far as to substitute artifice for nature, as if anything man could do was more beautiful than what nature did!" (p. 6). The idea of establishing a protected area focuses on the

> idea of conserving natural treasures, which are also the artistic and scientific heritage of the nation. It was desired to save it, preserving it, safeguarding it from new attacks, at least the little of the most beautiful works of nature that still remained intact.
>
> (pp. 7–8)

The idea that nature is a superior entity, endowed with a creative flair capable of generating aesthetic value, clearly emerges from these quotes.

It is important to note that, among the so-called natural elements whose protection is advocated, "picturesque localities" are listed together with rocks, fossils, caves, erratic boulders, flora, and fauna. Rural settlements, located within contexts defined as natural, since they cannot be removed and are the clear evidence of the work of human beings in these ecosystems, are described as results of the work of an abstract nature: the paradox is evident. Briefly, human action is described as real only when it is destructive, because when it is constructive and contributes to the functioning of ecosystems or the enhancement of their aesthetic value, is absorbed into natural action. In this transfer, an ethical problem also arises, since it is evident that these "picturesque localities" owe their appearance and function to the work of local communities, who are expropriated of their territorialising function, which has contributed to generate the ecosystems that the institution of the ANP aims at protecting.

The assonance with the rhetoric that accompanied the first national parks in the United States of America, extolling the wilderness, is therefore not surprising: "a primitive wilderness, a forest, one might say, still virgin, almost everywhere dense and thick, for not a few inextricable stretches, of a superb splendour, of an unsurpassable magnificence" (Pirotta, 1917, pp. 24–25). This is a critical point also because the discourses on nature were conveyed in the context of a nationalist rhetoric at the turn of the 20th century, as a part of the broader process of the construction of the Italian nation. The construction of a transcendental idea of nature, oblivious of the role of human communities in the historical interaction with ecosystems, generated a dichotomous vision, which led to the marginalisation of rural communities. The nationalist principles emerge in particular in some sections of the text as the following: "nothing urgent has yet been done in Italy, not even for the decorum and good name of Italy" (p. 10). The effect is the generation of a discourse that leads to the construction of an idea of nature associated with the essence of the landscapes with which Italians are expected to identify, based on the dichotomy between what is human and what is not.

The years that followed the foundation of the national park

A second document of crucial importance is the Sipari Report,[5] presented in 1926 by the MP Erminio Sipari, the first president of the ANP and one of the most important supporters of the need to fund it. This is a historical document that not only reconstructs the various stages of the process that led to Royal Decree in 1923, by which ANP was formally founded, but it also reveals a lot about the conditions of the territory from the ecological and anthropic points of view in the first two decades of the 20th century. However, this is not all, because some passages of this huge document (about three hundred pages long) testify an extraordinarily advanced vision about the protection of ecosystems, explicitly inspired by the example of Yellowstone National Park. It does not look only at naturalistic aspects, but is based on a vision that reconciles ecological and economic aspects, emphasising the potential of ANP as a tool for integrating territorial planning, ecology, and the wellbeing of local communities.

Obviously, it does not lack aspects on which it is possible to reason critically, especially from two points of view. First, the relationship with wildlife protection: one of the park's priorities from the earliest years has been the protection of the bear and chamois population, which are described in emphatic and celebratory tones, but also emphasising the need to educate the local population and the visitors on a conscious, peaceful, and responsible coexistence. Explicative of this approach are passages such as the following one:

> Nor, on the other hand, can the wildlife be disturbed by the development of tourism [...] all these voices from the mountains have been familiar to bears for centuries, and it will not be a few groups of cheerful tourists that will noticeably increase these noises [...] the bear defends itself: it is only necessary to remove the insidiousness of the firearm from it.
>
> (Sipari, 1926/2020, pp. 214–217)

In contrast to this advanced approach towards bear protection, the author's view of the wolf is contradictory: "therefore the wolf [...] is also the Park's enemy" (p. 223). It is evident that, reading these words today, we find it hard to agree with them, but we can understand this position considering that studies on the ethology of wolves consolidated in Italy only in the second half of the twentieth century. To understand this position on wildlife, we can delve into a second key aspect of the Sipari Report that deserves a critical reflection. Indeed, the fundamental importance of tourism as a key economic activity for the revitalisation of the territory is emphasised repeatedly. This is obviously not a problem, at that time tourism was beginning to emerge as a mass phenomenon, therefore its economic potential could not be underestimated. Sipari's attention to the actions necessary to transform the ANP into a tourist area is evident and to the point, reviewing all areas, from the road network to receptivity to seasonality. Also in this case, critics must be circumstantiated to the perspective of the author's viewpoint, without foreknowledge of the following hundred years of tourism history. However, we must emphasise that the idea of totally reconverting the rural economy into a tourist economy that seems to emerge has exposed the territory to critical issues connected to the abandonment of age-old practices of land use on which many ecological dynamics depend. At the same time, we can notice that a strong awareness of the author emerges with respect to the profound rural structure of the territory, which must be safeguarded together with ecological dynamics:

> because of the need to let the hundreds of thousands of our sheep (indispensable for the economy of our villages) enjoy our rich pastures, and because of the need for our municipalities to exploit their rich forests, the Abruzzo Park cannot be "untouched".
>
> (Sipari, 1926/2020, p. 213)

The idea of nature that underlies the work of the first President of ANP is therefore extremely advanced and already very close to the integrating paradigm, that aims to favour the coexistence between human activities and ecological processes, based on awareness of the fundamental ecological role of agricultural, pastoral, and forestry activities.

Sipari's Report concludes with a series of statements extolling the importance of ANP. Many are obviously to be interpreted in the light of the historical circumstance defined by the strengthening of the fascist government during the 1920s. However, at least one can be extrapolated as an example:

> You must also get to know the people of that region, you must approach them in order to appreciate the high qualities that derive from our most ancient, genuine and uncorrupted stock, the same stock that, many centuries before the foundation of Mother Rome, called itself Italica and gave Italy its name.
>
> (Sipari, 1926/2020, p. 262)

This quote shows how, also in Sipari's intentions, the protection of ANP's luxuriant fauna and the strengthening of the tourism sector were also functional to the construction of an identity discourse for the nascent Italian nation, which should have self-identified in the so-called "natural beauties" and could build through them a process of hetero-representation capable of making Italy recognisable in the world.

Discussion and conclusions

This chapter has aimed to highlight the role of PAs in the process of the social construction of nature, showing how they do not aim only to protect the ecological values of a region, nor do they complete their role as territorial entities aimed at fostering the synergy between antropic activities and ecological dynamics. Nor is their importance less significant as narrative tools that convey meanings related to the idea of nature and to the relationship between humans and non-humans, that can be either dichotomic or integrated. Obviously, radically different views of the world derive from one or the other perspective.

Testing the potential of the dialogue between geography and environmental history, this chapter has attempted to study the historical process that led, through the foundation of the first national park in Italy, to the construction of an idea of nature that contributed to the creation of Italian national identity. This is, of course, a complex process that involves many other factors (Banti, 2011; Thiesse, 1999), but the purpose is to underline the role of national parks within this process, as drivers for the social construction of a specific idea of nature on whom the relationship between human local communities, ecosystems, and living beings have been founded in Italy.

As has been shown through the cited historical sources, referring to natural elements (such as woods) as symbols of wild nature, is usual to construct an idea of nature associated with idealised aesthetic values. Despite this, in contexts such as Italy, it is precisely woods and other spaces usually considered as natural that provide the best opportunities to understand the prevalence of hybridisations that blur the boundaries between humans and non-humans. All Italian forests are indeed the result of the centuries-long constructive interaction between ecosystems and human beings, whose work has shaped their forms and ecological characteristics (Agnoletti, 2018).

Neither the idea of an aestheticised nature, nor the nihilist perspective that demonises the role of humans in ecosystems is helpful. Both of these perspectives are not based on objective factual data, but on ideological visions that matured between the 19th and 20th centuries. Questioning these perspectives today, also considering the awareness of their role in the construction of national identities, is crucial.

Coherently with the more recent critical ideas on nature conservation (Büscher & Fletcher, 2020), this chapter proposes to overcome the idea to study the relationships between human beings and ecosystems by excluding the political and economic variables that affect them. By studying the historical process of the construction of the idea of nature, we have attempted to show how it is closely correlated with the political project aimed at constructing a nation-state in Italy based on the economic principles of capitalism. In this specific case, they have led to the commodification of nature through tourism, indirectly exacerbating the separation between human and non-human entities. The need to address these problems is nowadays more urgent than ever, for practical and ethical reasons, which require humans to rethink their role in the ecumene (Berque, 1996), because Anthropocene is actually a fact and global change has definitively and inexorably snowballed its effects.

Notes

1 Nowadays the park is called Abruzzo Lazio and Molise National Park (ALMNP), however in this chapter I will use the term Abruzzo National Park (ANP), as I refer to the historical period of its foundation, when the park's extension and thus its name did not include the neighbouring regions of Lazio and Molise.
2 Cited in Gentile 2010, p. 16. All translations are by the author.
3 Giovanni Rosadi, *Per la difesa del paesaggio*, Tipografia Bonducciana, 1911. Quoted in Piccioni, 2014, p. 166.
4 See L. n. 778, 11/06/1922.
5 In this chapter I am analysing this document only in its narrative and symbolic function, but I want to point out that it also contains an overall presentation of the practical actions carried out by the park in the first years after its establishment (related to the protection of flora and fauna, forest management, accessibility, tourism, etc.), which I cannot explore in depth for reasons of space.

References

Agnoletti, M. (2018). *Storia del bosco. Il paesaggio forestale italiano*. Laterza.

Armiero, M. (2022). *La natura del duce. Una storia ambientale del fascismo*. Torino.

Armiero, M., & Barca, S. (2004). *Storia dell'ambiente. Una introduzione*. Carocci.

Arneth, A., Leadley, P., Claudet, J., Coll, M., Rondinini, C., Rounsevell, M. D. A., Shin, Y.-J., Alexander, P., & Fuchs, R. (2023). Making protected areas effective for biodiversity, climate and food. *Global Change Biology*, *29*, 3883–3894. https://doi.org/10.1111/gcb.16664

Banti, A. M. (2011). *Sublime madre nostra. La nazione italiana dal Risorgimento al fascismo*, Laterza.

Berque, A. (1996). *Etre humains sur la Terre. Principes d'éthiques de l'écoumène*. Gallimard.

Büscher, B., & Fletcher, R. (2020). *The conservation revolution: Radical ideas for saving nature beyond the Anthropocene*. Verso.

Castree, N. (2014). *Making sense of nature. Representation, politics and democracy*. Routledge.

Castree, N., & Braun, B. (2001). *Social nature: Theory, practice, and politics*. Wiley-Blackwell.

Cronon, W. (1996). The trouble with wilderness: Or, getting back to the wrong nature. *Environmental History*, *1*(1), 7–28. https://doi.org/10.2307/3985059

Demeritt, D. (2002). What is the 'social construction of nature'? A typology and sympathetic critique. *Progress in Human Geography*, *26*(6), 767–790. https://doi.org/10.1191/0309132502ph402oa

Depraz, S. (2008). *Géographie des espaces naturels protégés. Genèse, principes et enjeux territoriaux*. Armand Colin.

Ferrari, C. (2015). Alessandro Ghigi e la Società Emiliana Pro Montibus et Sylvis. *Natura e montagna*, *LXII*(3), 30–36.

Gentile, E. (2010). *Né stato né nazione italiani senza meta*. Laterza.

Giacomini, V., & Romani, V. (1982). *Uomini e parchi*. Franco Angeli.

Hardenberg, W., Kelly, M., Leal, C., & Wakild, E. (Eds.). (2019). *The nature state. Rethinking the history of conservation*. Routledge.

Moschini, R. (2013). *Parchi e politica*. ETS.
Nash, R. F. (2014). *Wilderness and the American mind*. Yale University Press (original work published 1967).
Pedrotti, F. (2012). *I pionieri della protezione della natura in Italia*. Temi.
Phillips, A. (2003). Turning ideas on their head. The new paradigm for protected areas. *The George Wright FORUM*, *20*(2), 8–32.
Piccioni, L. (2014). *Il volto amato della patria. Il primo movimento per la protezione della natura in Italia 1880–1934*. Temi.
Piccioni, L. (2023). *Parchi naturali. Storia delle aree protette in Italia*. Il Mulino.
Pirotta, R., (1917). *Il Parco Nazionale dell'Abruzzo*. Federazione italiana delle associazioni Pro Montibus ed enti affini.
Sievert, J. (2000). *The origins of nature conservation in Italy*. Peter Lang.
Sipari, E. (2020). *Relazione Sipari*. Edizioni del Parco Nazionale d'Abruzzo, Lazio e Molise (original work published 1926).
Smith, N. (2008). *Uneven development: Nature. Capital, and the production of space* (3rd edition). The University of Georgia Press (original work published 1984).
Soriani, S., Vallerani, F., & Zanetto, G. (Eds.). (1996). *Nature, environment, landscape: European attitudes and discourses in the modern period, the Italian case 1920–1970*. Quaderni del Dipartimento di Geografia (18), Università di Padova.
Thiesse, A. M. (1999). *La Création des identités nationales Europe, XVIIIe-XXe siècle*. Seuil.
Thomas, L., Middleton, J., & Phillips, A. (2003). *Guidelines for management planning of protected areas*. IUCN.
Tuan, Y. F. (1990). *Topophilia. A study of environmental perception, attitudes, and values*. Columbia University Press (original work published 1974).
Turco, A. (2010). *Configurazioni della territorialità*. Franco Angeli.
Vidali, P. (2022). *Storia dell'idea di natura. Dal pensiero greco alla coscienza dell'antropocene*. Mimesis.
Whatmore, S. (2002). *Hybrid geographies. Natures cultures spaces*. Sage.
Zanolin, G., & Paül, V. (2022). Exploring the sustainability of wilderness narratives in Europe: Reflections from Val Grande National park (Italy), *Geographical Review*, *112*(3), 444–465. https://doi.org/10.1080/00167428.2020.1869905
Zimmerer, K. S. (2000). The reworking of conservation geographies: Nonequilibrium landscapes and nature-society hybrids. *Annals of the Association of American Geographers*, *90*(2), 356–369. https://doi.org/10.1111/0004-5608.00199

2 Transcending nature conservation in Italian UNESCO biosphere reserves

Margherita Cisani

UNESCO's Man and the Biosphere programme and the debates on conservation and landscape

UNESCO's biosphere reserves (BRs), as sites of research and action for biological and cultural conservation, were created as part of the intergovernmental scientific programme Man and the Biosphere (MAB). They were launched by UNESCO in 1971 to promote a balanced relationship between man[1] and the environment through the protection of biodiversity. The programme, strongly supported by Italian scientists (including Francesco Di Castri, who was the first secretary of the MAB programme), sees BRs as living laboratories where sustainable coexistence between living communities in complex ecosystems can be experimented and promoted. The World Network of Biosphere Reserves, although not as renowned as the list of World Heritage Sites, currently includes 759 sites in 136 countries globally, 21 of which are in Italy.

Although resembling protected areas in some ways, with biodiversity conservation as their primary objective, BRs were created with the idea of combining nature protection with the enhancement and sustainable development of local communities. They are conceived and promoted as territorial laboratories to foster a lasting balance between the needs for conservation of biological and cultural diversity and socio-economic development, and represent 'the most longstanding, organized, international, and intergovernmental network to support the aims of sustainability science' (Reed & Price, 2020, p. 2).

In the Statutory Framework, a further definition of BR is offered as 'sites of excellence to explore and demonstrate approaches to conservation and sustainable development on a regional scale' (UNESCO, 2020, art. 3), and the following three constitutive functions are specified: conservation of biodiversity and cultural diversity, with specific reference to landscapes, ecosystems, species and genetic diversity; economic and human development that is socio-culturally and ecologically sustainable; and logistical support for development through demonstration projects, environmental education, training, research and monitoring of local, regional, national and global issues and themes related to conservation and sustainable development.

DOI: 10.4324/9781003465164-4

On the basis of the abovementioned functions, BRs are internally divided into three distinct zones: core, buffer and transition. Core zones, consistent with the objectives of a BR, are legally protected areas dedicated to long-term preservation and of a sufficient size to guarantee the achievement of the objectives. Within these zones, only conservation and research activities are permitted. Buffer zones surround or border core areas and are only used for activities that are compatible with the conservation objectives and ecological practices to enhance scientific research, monitoring, training and education. Transition zones are conceived as inhabited areas where communities promote economic development activities that are socio-culturally and environmentally sustainable.

The fact that, from the very beginning, BRs have been promoted as sites for scientific research into a sustainable relationship between humans and nature places them in connection with at least three major academic debates straddling the natural sciences, the humanities and social sciences: the debate on the social construction of nature (Castree & Braun, 2001), the discussion around the 'great conservation debate' (Büscher & Fletcher, 2020) and different meanings of the concept of landscape (Wylie, 2007).

From the viewpoint of social nature theory, BRs can be understood as an attempt, different from protected areas, to rework the boundaries of what we consider nature by responding to the challenges of the Anthropocene. This is achieved, on the one hand, by bringing nature closer to living contexts and, on the other hand, by inserting it into the global circuits of value creation and exchange, be it a patrimonial value (through forms of ecotourism and cultural tourism) or an ecological value (through ecosystem services or carbon credit exchange). Therefore, it is necessary to frame BRs within what Büscher and Fletcher called the mechanism of 'accumulation by conservation'. According to them, the institution of BRs is a characteristic instrument of the second phase of global conservation development, which corresponds to neoliberal, decentralised and flexible conservation (Büscher & Fletcher, 2015, p. 13).

Finally, BRs are an interesting case study to observe how landscape-as-heritage making processes are implemented; that is, how the recognition of a patrimonial value to the landscape, or to some of its characteristics or elements, occurs and with what consequences (Harvey, 2015; Pettenati, 2023). BRs speak, explicitly or not, about landscape, drawing combined attention towards ecological and socio-cultural values and adopting a diachronic vision that can overcome the dichotomy between conservation of traditions and sustainable innovation (Aschenbrand & Michler, 2021).

Research context and methodology

This chapter examines the entire Italian territory, considered globally as a hotspot for both biodiversity conservation and the protection of historical, cultural and landscape heritage. In this context, a double bias can be observed. On the one hand, BRs are one of the many contexts for studying the issues related to the conservation of animal and plant biodiversity or geosciences, therefore, while there is extensive research on their ecological aspects, only a few studies have considered them from a socio-economic and geographical perspective (Di Lonardo & Cinocca, 2021). On the other hand, the Italian debate on landscape conservation is strongly influenced by a prevalent aesthetic approach, often focused on preserving the cultural, historical and artistic values of landscape (Cassatella & Castiglioni, 2019), largely ignoring the meaning and the role of BRs. This chapter, therefore, aims to fill these gaps in the research on Italian BRs, for which there is not even an institutional public cartography, by analysing them from a geographical viewpoint.

Italy, with 21 BRs, currently ranks sixth among the countries with the maximum number of BRs. This list is headed by Spain, where the network is particularly well-developed but shows problematic aspects and general disengagement from the local arena (Paül et al., 2022, p. 9). Paül et al. argued that BRs are 'simply political and economic products that have little in common with the theoretical environmental and sustainable development dimensions described in the UNESCO framework on BRs' (2022, p. 8). Therefore, the conspicuous number of BRs in Italy, rather than being taken simply as a success, must lead to reflection on the reasons and the modalities with which such initiatives are spatialised and interact with local socio-environmental dynamics.

A final important aspect to be considered, among the reasons for the interest in the Italian case study, is the fact that BRs are also posing themselves as tools for managing the ever-increasing pressure of tourism (and not only) towards natural areas close to or embedded in inhabited contexts. This is the result of the growing importance placed on nature and biodiversity, and therefore on BRs, as a source of wellbeing, which manifests itself in the growth of slow and proximity ecotourism, in Italy as elsewhere, especially in the post-pandemic era (Mondino & Beery, 2019).

This study was based on the collection of cartographic data, which were then analysed using GIS software, and on the collection and analysis of official documents (mainly application dossiers and action plans). Note that, within the official websites of the individual reserves, let alone the pages dedicated to the Italian network within the Ministry of the Environment (MASE) portal, this information is not present and the Italian BRs are only briefly described, indicating the establishment year. Some information was collected through the World Network of Biosphere Reserves website, which also lists the contact details of the managing body. However, most of the information for each reserve was collected through emails or telephone contacts. Further information was collected through research on each BR's official website and social media. Therefore, the results presented here come mainly from these sources. However, the research also relied on the qualitative in-depth analysis of specific case studies, which will be briefly presented here and whose analysis contributed to shaping the reflections exposed here. In particular, this chapter aims to reveal the various ways in which BRs are transcending the notion of nature conservation through landscape-as-heritage making, addressing the following research questions: (i) Where are the BRs in Italy located? (ii) What narratives about landscape and nature do they (re)produce? (iii) What kind of initiatives and projects related to natural and cultural heritage do they promote? To answer the first question, this chapter proceeds with the presentation of the Italian network, followed by an analysis of the BRs narratives and projects, aimed at answering the second and third questions.

The Italian network of BRs

As indicated above, there are 21 BRs in Italy (see Figure 2.1), two of which are transboundary (i.e., located in border territories) and therefore managed in a coordinated manner with homologous bodies on transalpine territories. These are the Mont-Viso Transboundary Biosphere Reserve, between France and Italy, and Julian Alps, between Italy and Slovenia. The BRs are located in 14 regions (13 regions, 1 autonomous province), mainly in the northern part of the country, and comprise 856 municipalities, which are inhabited by approximately 6 million people. The total area within the BRs corresponds to 38,792 km^2, roughly 1/10th of the entire country, and the vast majority is located in mountain territories (16 out of 20 BRs include mountain areas above 600 m.a.s.l.).

On average, BRs occupy an area of almost 2,000 km^2—varying from 30 km^2 of Miramare to almost 11,000 km^2 of the Tuscan Islands—distributed among the core area (5%), buffer area (19%), and transition area (76%). Of these, the transition zone distinguishes the BRs to the maximum extent, since compared to the traditional zoning within protected areas, it adds a further level of involvement intended to include infrastructures and densely populated areas in the designation. However, the transition zone is also the most difficult to define and put into practice due to the complexity and fragmentation of its governance, which involves different actors who must be able to share the objectives of conservation, development and logistical support. Communication and spreading awareness about the significance of the transition zone are also complex aspects. In fact, it is not uncommon for inhabitants included in the transition zone to be unaware that they are part of a UNESCO BR.

All BRs overlap with some form of protected area, since by definition, the core area must be preserved through legislative instruments. In the Italian context, we find the presence of 11 national parks, 3 marine protected areas, 60 regional parks or reserves and 13 other protected areas within the boundaries of the BRs. As in any heritage-making process, setting the boundaries and overlapping of the site is a complex process of negotiation between different actors, a process that leads to the selection of some areas to be included while inevitably excluding others. This occurs on the basis of the encounter between local instances and UNESCO's globalising logics (the requirements of the BR Statutory Framework) and leads to the selection of areas that correspond the most to the idea of territory that the BR wants to represent, which is never univocal and inevitably excludes and overshadows other possible representations (Pettenati, 2019).

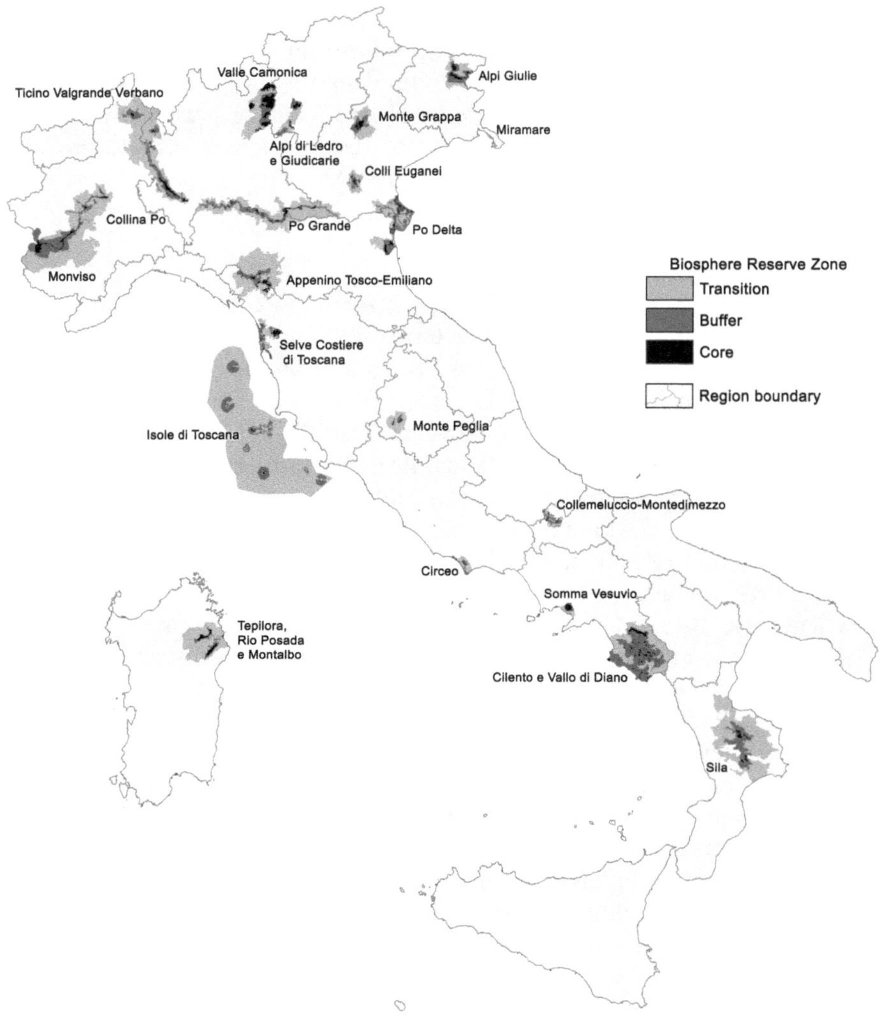

Figure 2.1 Map of the Italian BR boundaries and zoning. Cartographic elaboration by Francesco Ferrarese on data collected by the author.

Therefore, the nomination process is critical to the analysis. According to Italian guidelines, it follows a series of passages among the local, national and global scales. First, any local authority, association or other proponents activates the process by contacting the MASE national committee. Within roughly a couple of years of local and national negotiations, the nomination dossier is produced following the MAB UNESCO Statutory Framework indications and responding explicitly to seven criteria that assess the proposed BR's ability to respond to the three constitutive functions with an adequate zoning and with effective managing structures and participatory processes. The International

Advisory Committee for Biosphere Reserves then examines the proposal and recommends to the MAB Council, usually meeting once a year, either to approve it or to defer it.

Italian BRs are not officially recognised under a nature protection law, as the only official document that regulates them are the national guidelines provided by the MASE, which is also the seat of the national technical committee that liaises with UNESCO, manages national nominations and supports local authorities in implementing the MAB programme. Therefore, management bodies differ from case to case, from national or regional parks to consortia, municipalities and other local institutions (Table 2.1).

Table 2.1 Italian BRs. elaboration by the author on data from UNESCO, MASE and official BR websites

BR name	*Designation year (update year)*	*Management institution*
Circeo	1977 (2012)	Parco Nazionale del Circeo
Collemeluccio-Montedimezzo	1977 (2012)	Consorzio AssoMAB Alto Molise
Miramare	1979 (1999, 2011)	Area Marina Protetta Miramare
Cilento e Vallo di Diano	1997 (2013)	Parco Nazionale del Cilento
Somma-Vesuvio e Miglio d'Oro	1997 (2013)	Parco Nazionale del Vesuvio
Ticino, Val Grande Verbano	2002 (2013)	Parco Lombardo della Valle del Ticino
Isole di Toscana	2003 (2014)	Parco Nazionale Arcipelago Toscano
Selve costiere di Toscana	2004 (2016)	Parco regionale Migliarino San Rossore Massaciuccoli
Monviso	2013	Ente di Gestione delle Aree Protette del Monviso
Sila	2014	Parco Nazionale della Sila
Appennino Tosco-Emiliano	2015 (2021)	Parco Nazionale Appennino Tosco Emiliano
Alpi Ledrensi e Judicaria	2015	Consorzio dei comuni B.I.M. Sarca Mincio Garda
Delta del Po	2015	Parco Regionale Delta del Po
Collina Po	2016	Ente di Gestione delle Aree Protette del Po e della Collina
Tepilora, Rio Posada e Montalbo	2017	Parco Naturale Regionale di Tepilora
Valle Camonica-Alto Sebino	2018	Comunità Montana di Valle Camonica
Monte Peglia	2018	Associazione Monte Peglia per Unesco
Po Grande	2019	Autorità di Bacino Distrettuale del fiume Po
Alpi Giulie	2019	Parco Naturale Regionale delle Prealpi Giulie
Monte Grappa	2021	IPA Asolo Monte Grappa
Colli Euganei	2024	Parco Regionale dei Colli Euganei

The first Italian BRs date back to the 1970s and concern areas whose conservation is aimed at maintaining and safeguarding their biodiversity. The first generation of Italian BRs thus comprises the three BRs of Collemeluccio-Montedimezzo in Molise, Circeo in Lazio and Miramare in Friuli Venezia Giulia. From the so-called Seville Action Plan (i.e., from 1996 onwards), two BRs were established in 1997 and three between 2002 and 2004. Following the Madrid Action Plan (UNESCO, 2008), which strongly emphasised the role of BRs in countering three major challenges of the millennium (climate change, accelerated biodiversity loss and urbanisation), Italy saw five more BRs recognised, to which the seven most recent BRs were then added following the Lima Action Plan (UNESCO, 2016).

The fact that there is no reference legislation in Italy means that, in practice, there is an extreme diversity of objectives, dimensions, actors and management models. In addition, the prominent role of the MASE, compared to other potentially involved ministries, means that an approach linked to the conservation of ecosystems and natural heritage prevails instead of a real overcoming of the dichotomy between natural and cultural heritage.

The MASE national guidelines require BRs to meet at least once a year for a national meeting, with the aim of comparing their activities and achievements, exchanging good practices, addressing any problems encountered and promoting networking activities between reserves, both nationally and internationally. A Memorandum of Understanding was signed in November 2021, where the Italian BRs committed to set up a coordination committee to define meetings and activate working groups on issues identified as being of particular interest. These issues included external and internal communication of the values of the MAB programme and the role of BRs; involvement of stakeholders and the private sector; participation of young people; contribution of the reserve network to the creation of a database in collaboration with the MAB National Technical Committee; and effective collaboration with the National Technical Committee and foreign national associations of other BRs, as well as Italian or foreign networks of other UNESCO recognitions. These issues reflect well the already identified gaps, as well as the contradictions and trends in the different ways in which BRs reproduce and elaborate on the natural and cultural heritage they represent and seek to conserve and enhance.

Re-producing nature in BRs

Following Aschenbrand and Michler (2021), we can interpret the notoriety gap between BRs and national parks based on the different narratives they offer (i.e., the different discourses these institutions refer to in the way they present themselves to the public). The analysis conducted in Germany indicated that the 'cultural landscape' narrative is the one that is most used to highlight the importance of traditional rural landscapes, handcrafts and locally produced food, while national parks refer more to the narrative anchored in the concept of wilderness. Aschenbrand (2022) argued, through the concept of extended urbanisation, that highlighting the links between urban and rural development could help dissolve the dichotomy. Considering BRs, in particular, this concept is also useful in reconciling the functions of conservation and development.

Thus, the ways in which BRs attempt to dissolve the nature–culture dichotomy are multiple and closely related to the narratives and discourses of nature that they embody and reproduce. In the final analysis, BRs can be interpreted as an instrument for the reconfiguration and reproduction of nature that particularly leverages a notion of value extended not only to the ecological but also to the cultural, social and economic spheres.

From the research conducted on the nomination documents of Italian BRs, as well as on the websites and official descriptions, Aschenbrand and Michler's interpretation can be partially confirmed also with regard to Italian BRs. The term 'landscape', in fact, frequently appears in numerous dossiers (16 dossiers out of 21), while terms such as species, habitat, park, forest and plain prevail in the descriptions. In these instances, landscape is frequently understood as witnessing an alleged harmony between natural and socio-cultural values, often only juxtaposed rather than integrated. In fact, there are numerous references to traditional agricultural landscapes or to the landscape as a beautiful setting for recreational activities.

The notion of cultural landscape as heritage is, therefore, certainly present within the narratives of Italian BRs; however, the dichotomy between nature and culture is not completely dissolved, as these values are often simply juxtaposed and not integrated. In a way, however, BRs do play into the contradiction and, compared to other cultural or natural heritage instruments, explicitly inhabit the tensions between nature and culture, between tradition and innovation and between human and non-human.

The cultural landscape narrative linked to tradition and rurality is reflected in the type of activities that Italian BRs propose and which, net of the initiatives aimed at studying and monitoring useful for the conservation of biodiversity in the strict sense, often concern activities for the promotion of traditional agricultural practices, wine and food products or slow tourism, such as walks, paths and itineraries for cycling tourism, which connects natural sites to those of cultural interest. Examples include UPVIVIUM, the food and wine contest of the Italian MAB UNESCO BRs, or buying groups, alternative food networks, markets, fairs, shops, courses, tastings and workshops dedicated to organic farming in BRs. In other cases, the BRs promote trekking itineraries led by experts in the area for the local population, as well as for visitors, with the objective of combining the promotion of sustainable tourism with raising public awareness of the values of the reserves. In some cases, these treks are also linked to the presentations of books or documentaries and see the mixed participation of experts and laymen, promoting convivial and horizontal exchange of knowledge without envisaging the production of economic value. In contrast, some reserves fully embrace the 'selling nature to save it' approach, as in the case of Appennino Tosco Emiliano, where a local system of remuneration for ecosystem services generated by forests has been introduced through the coordinated and wide-area exercise of sustainable and responsible forest management. The same BR, however, as proof of the complex contradictions it embodies, supports the creation of community enterprises, cooperatives of inhabitants that collectively manage not only the production of goods but also the landscape and collective services, such as transports, welfare or education.

At first sight, landscape in BRs seems to be understood mainly as an object to be conserved or promoted. However, there are also experiences that see the development of a deeper reflection on landscape as an integrated natural and cultural heritage and that try to engage with and educate residents and visitors about the complex nature–culture nexus represented by the BRs. This is done precisely by using landscape as a tool for local sustainable development, taking into account socio-cultural aspects and involving research centres, archives, libraries and universities.

Conclusions

The analysis of Italian BRs revealed numerous contradictions inherent in the MAB programme itself. Within these contradictions, the BRs seem to simultaneously propose different ways of going beyond mere nature conservation. Based on the reflections reported at the beginning, at least three of them have been highlighted in this chapter: the transcendence of conservation without integration, the transcendence of conservation into development and the transcendence of conservation into conviviality.

The transcendence of conservation without integration is essentially a negation of the guiding principles of the MAB programme but is evident in the fact that conservation in the BRs is mainly concentrated in the core zones and oriented towards a traditional protection of resources, ecosystems and animal and plant species. In contrast, the development function, which is constitutive of the BRs, is concentrated in the buffer zones and especially in the transition zone. Here, natural heritage is complemented by cultural heritage as a lever for economic development, especially of the tourism–recreational type.

The second type, transcendence into development, is well represented by the proliferation of economic instruments aimed at the exploitation of protected nature within the framework of market mechanisms. This indicates that BRs are fully integrated into the 'mainstream conservation' model, which opens up to a less dichotomous view (bringing conservation closer to the people) but continues to rely on capitalist mechanisms (accumulation through conservation) to support nature conservation.

Finally, the last type envisages a paradigm shift from conservation to conviviality, as described by Büscher and Fletcher (2020). It includes experiences that promote forms of coexistence between humans and non-humans within protected areas, which favour the joint management of resources beyond market mechanisms through cooperative forms or by simply emphasising the cultural and non-profit initiatives of awareness, engagement and participation in heritage making.

In the Italian context, the research conducted and presented in this chapter also shows that newer BRs are more often managed by institutions other than protected areas, such as other local administrations, consortia or associations. This could be evidence of a gradual shift from conservation-oriented management to more integrated management. Further research is needed to understand whether this more integrated approach is solely aimed at economic development (albeit sustainable) or whether it is possible to see in BRs an attempt to transcend mainstream conservation beyond the nature–culture dichotomy in a post-capitalist way.

Note

1 The MAB programme is being increasingly critiqued for its widespread use of "man" instead of "humans" or "mankind", and multiple attempts have been made to change the official denomination of the programme. Despite this, during the most recent 36th session of the MAB Council, held in Agadir (Morocco), parties voted against any proposal of change in denomination.

References

Aschenbrand, E. (2022). How can we promote sustainable regional development and biodiversity conservation in regions with demographic decline? The case of UNESCO biosphere reserve Elbe River Landscape Brandenburg, Germany. *Land*, *11*(10), Article 1623. https://doi.org/10.3390/land11101623

Aschenbrand, E., & Michler, T. (2021). Why do UNESCO biosphere reserves get less recognition than national parks? A landscape research perspective on protected area narratives in Germany. *Sustainability*, *13*(24), Article 13647. https://doi.org/10.3390/su132413647

Büscher, B., & Fletcher, R. (2015). Accumulation by conservation. *New Political Economy*, *20*(2), 273–298. https://doi.org/10.1080/13563467.2014.923824

Büscher, B., & Fletcher, R. (2020). *The conservation revolution: Radical ideas for saving nature beyond the Anthropocene*. Verso Books.

Cassatella, C., & Castiglioni, B. (2019). Protection, management and planning: Ragioni, pratiche ed effetti della tutela del paesaggio. *Ri-Vista. Research for Landscape Architecture*, *17*(2), 5–13. https://doi.org/10.13128/rv-8308

Castree, N., & Braun, B. (Eds.). (2001). *Social nature: Theory, practice, and politics*. Blackwell Publishers.

Di Lonardo, S., & Cinocca, A. (2021). A man and the biosphere reserve as a natural and socio-economic laboratory for the sustainable future of small rural communities. *Eco.Mont (Journal on Protected Mountain Areas Research)*, *13*(special issue), 125–129. https://doi.org/10.1553/eco.mont-13-sis125

Harvey, D. (2015). Landscape and heritage: Trajectories and consequences. *Landscape Research*, *40*(8), 911–924. https://doi.org/10.1080/01426397.2014.967668

Mondino, E., & Beery, T. (2019). Ecotourism as a learning tool for sustainable development. The case of Monviso Transboundary Biosphere Reserve, Italy. *Journal of Ecotourism*, *18*(2), 107–121. https://doi.org/10.1080/14724049.2018.1462371

Paül, V., Vila-Lage, R., & Trillo-Santamaría, J.-M. (2022). "The n°1 country"? A critical investigation of the booming designation of biosphere reserves in Spain. *Landscape and Urban Planning*, *222*, Article 104375. https://doi.org/10.1016/j.landurbplan.2022.104375

Pettenati, G. (2019). *I Paesaggi culturali Unesco in Italia*. FrancoAngeli.

Pettenati, G. (2023). *Landscape as heritage: International critical perspectives*. Taylor & Francis Group.

Reed, M. G., & Price, M. F. (2020). *UNESCO biosphere reserves: Supporting biocultural diversity, sustainability and society*. Routledge.

UNESCO. (2008). Madrid action plan for biosphere reserves (2008–2013). https://unesdoc.unesco.org/ark:/48223/pf0000163301?posInSet=24&queryId=28164174-4e57-42ae-8049-0437c28838ee

UNESCO. (2016). Lima action plan for UNESCO's man and the biosphere (MAB) programme and its world network of biosphere reserves (2016–2025). https://unesdoc.unesco.org/ark:/48223/pf0000381215

UNESCO. (2020). *Statutory framework of the world network of biosphere reserves*. https://unesdoc.unesco.org/ark:/48223/pf0000373378

Wylie, J. (2007). *Landscape*. Routledge. https://doi.org/10.4324/9780203480168

3 Rewilding natures in the Ligurian mountains

Pietro Piana, Lorenzo Brocada, Rebekka Dossche, and Enrico Priarone

Introduction

In the last decades, many parts of the Italian peninsula underwent large landscape changes linked to spontaneous rewilding[1] processes due to the abandonment of the traditional socio-economic agro-silvo-pastoral system (Moreno, 1990). While the woodland cover in the pre-industrial period was largely managed as coppiced or wooded pasture, today most of it is no longer regularly mantained. From 1936, the date of the first consistent mapping of forests at national scale, the wooded surface increased exponentially, and today more than a third of the Italian territory is covered by woodland (Agnoletti et al., 2022; Pallotta et al., 2022). This revegetation process does not mean a decrease of urban surfaces; instead, it is mostly dominated by the abandonment of previously managed woodland and the succession of secondary vegetation on former arable land and open areas (Pincin, 2020; Munafò, 2022).

Rewilding natures are a pervasive phenomenon in Liguria; the region is Italy's third smallest one (5,418 km^2) and is a narrow and long strip of land arching around the Mediterranean and stretching from the Italian French border to Tuscany for a length of 350 km. It is mostly characterised by mountainous (65%) and hilly terrain (35%), intersected by very few plain areas along the main riverbeds. Given the morphology of the region, the population is not evenly distributed: the coastline is heavily urbanised while the inner part is generally underpopulated. Only the main valley floors of the hinterland, particularly around Savona and Genoa, still have a moderate to high degree of urbanisation. As a result of a combination of economic, political, technological and socio-cultural processes in the twentieth century, today Liguria's natures show a distinctive dichotomy between coastal and mountain areas. The former is associated to a classic, somehow stereotypical image of Ligurian landscapes linked to sea tourism where nature commodification processes include land take phenomena, beaches privatisation and artificialisation of coasts and riverbeds. On the other hand, inland areas were characterised by intense human presence in the pre-industrial period and high landscape diversification, but from the post-war period they underwent a fast and apparently irreversible

DOI: 10.4324/9781003465164-5

spontaneous rewilding process, that resulted in the transition from a managed, cultural landscape to a post-cultural re-naturalised one (Cevasco, 2007).

Drawing upon a diverse range of sources (historical and modern cartography, census data, media analysis and interviews[2]), this chapter explores rewilding processes in Liguria. It first aims at analysing the phenomenon at regional scale with particular reference to forest growth. It then focuses on two case studies in the Ligurian Alps and Apennines, quantifying the loss of agricultural surface between 1929 and today. The choice of these areas was based on three criteria: they are both protected by regional natural parks (Alpi Liguri and Antola); they are located at the two extremes of the region and in two different mountain chains; they fall, respectively, in the Tyrrhenian and Po Valley catchments. The final section discusses challenges and opportunities of rewilding in Liguria, emphasising the need of a multilayered approach to landscape management that combines the preservation of traditional cultural landscapes, and the strengthening of nature-based economies linked to rewilding processes.

Landscape transformation and perception of rewilding processes in Liguria

In Liguria land abandonment processes and consequent rewilding has had a significant impact not only on the landscape itself, but also on its actors, human and non-human. Historical ecology literature has shown how the Ligurian interior underwent major changes with several socio-economic and physical geographical implications (Balzaretti et al., 2004; Cevasco, 2007; Hearn et al., 2014; Moreno, 1990; Rota, 1991). The abandonment of agro-silvo-pastoral practices has not only led to a growth of secondary vegetation and the progressive reduction of open areas, but also to the return of animals, particularly large mammals such as the wolf, the wild boar and deer that have recolonised rewilded spaces (Hearn et al., 2014; Piana et al., 2024).

Lack of maintenance caused a general deterioration of hydrogeological efficiency in an already vulnerable territory, due to the abandonment of slopes and traditional irrigation systems (Faccini et al., 2016). This is particularly true for terraced areas sustained by dry stone walls, which is by far the most quintessential element of 'traditional' Ligurian landscapes. Many projects aimed at their conservation and renovation have taken place in areas characterised by high tourism presence, mediatic attention and landscape commodification. These include Cinque Terre (included in the World Heritage List as cultural landscape) and Portofino, where, alongside the role in maintaining hydrogeological stability, it is undeniable how their conservation is also aimed at the image-creation of Ligurian 'domestic' natures.

But these projects only involve a minor part of the region, as most of terraced areas, particularly in the interior, have been erased by vegetation growth alongside many other elements of the traditional cultural landscape including mills, chestnut dryers, religious and military buildings. More recently vegetation growth took place in peri-urban and urban wastelands, generating unusual habitats and ecological niches for wildlife with increased human-animal interaction (Piana et al., 2024).

In 2015 Liguria was Italy's region with the highest woodland index: 71.5% = 387,244 ha woodland surface on the total regional surface, against the 36.7% of the national average (Gasparini et al., 2022). The percentage increases to 78% if we include 'evolving shrubs',[3] which is often the vegetation stage before woodland (Brocada et al., 2023). Liguria has been characterised by a high area of woodland even in the pre-industrial period (305,291 ha in 1936), with two of the four provinces as the most wooded in Italy: Genoa (>50%) and Savona (±60–65%) (Brengola, 1939). However, whilst in 1936 woodlands were largely managed, most of the forests today are abandoned. Although not as sharply as in other regions, the woodland cover in Liguria has increased through time: +22.6% between 1936 and 1985 and +26.8% between 1936 and 2015 (Ferretti et al., 2018; Gasparini et al., 2022). Figure 3.1 illustrates the exponential growth of woodland between 1936 and 1990, and a more stable situation between 1990 and 2018. The three maps were created unifying different specific tree categories under a more general category of 'woods'.

Today the increase of wooded areas is generally well received by the public opinion although such an attitude is not always justified by a real knowledge of the situation. In this sense, the success of communication campaigns and initiatives aimed at raising awareness on the problem of deforestation worldwide (where this is undeniably a serious issue) and a superficial communication by medias might have indirectly contributed to a distorted perception of the state of nature and woodland at local level. In 2022, the Italian Ministry of Ecological Transition shared data on forestation where Genoa was the Italian metropolitan city[4] with the highest wooded surface and the highest rate of green area/inhabitant (105 m^2 vs an average of 34 m^2). The news was positively received by the public opinion, institutions[5] and several media, such as the local newspaper *Il Secolo XIX* (2022). In particular, the latter interpreted it as a 'virtuous green record' neglecting that reforestation processes are not directly linked to a well-defined political strategy but to abandonment and that such phenomena do not involve the built-up area, where more green spaces would be necessary. The overall increase in vegetation has not had the beneficial effects usually attributed to it by public opinion and by a certain narrative perpetrated by public institutions. Green areas are less accessible due to the expansion of brambles and invasive vegetation in general, and lack of maintenance of the trail network. Contemporarily, biodiversity has not increased, as only a few predominant species take over, whilst many animals and plants connected to open areas (e.g., birds protected by European Birds and Habitat directives) see their habitat reduced due to the decrease of open surfaces (cf. Directive 2009/147/EC on the conservation of wild birds). The risk of fires has increased, as lack of forest maintenance leads to an accumulation of incendiary material (Tonini et al., 2020), especially in the case of conifers. Moreover, the so-called 'heat islands' in cities have not decreased, as trees have risen mainly on the edge of the built-up area.

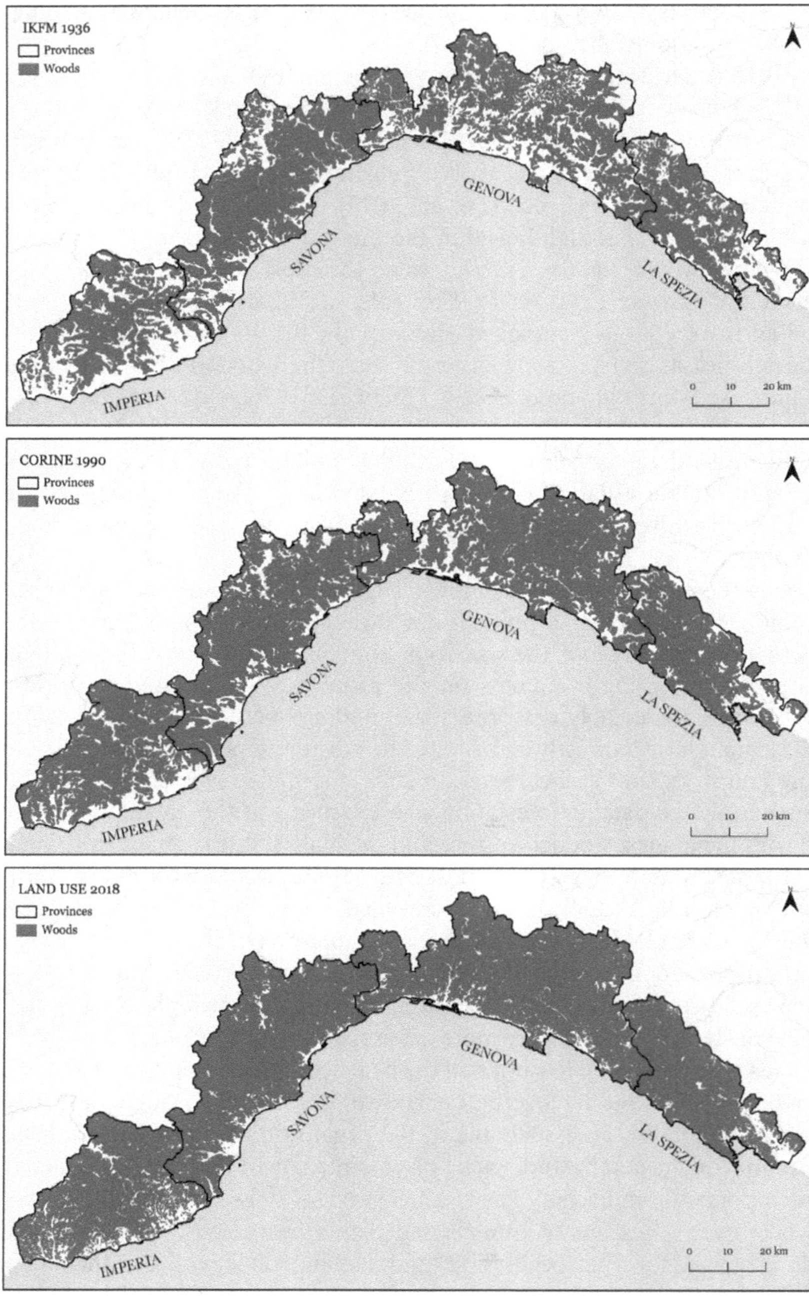

Figure 3.1 The evolution of woodland cover in Liguria between 1936 (Italian Kingdom Forest Map), 1990 (CORINE Land Cover) and 2018 (Ligurian Land Use Map). Made by the authors with QGIS.

Sources: Ferretti et al., 2018; Copernicus, 1990; Regione Liguria, 2018; ESRI.

Loss of agricultural activities – 2 Ligurian case studies

The increase of woodland during the last centuries is intrinsically related with the loss of agricultural activities. One of the best indicators to illustrate this development is the amount of surface used for agriculture collected through national census campaigns. This section quantifies the loss of Utilised Agricultural Area (UAA) during the same timespan as the map of Figure 3.1 (1936–1990–2018). Comparative analysis of the woodland cover in combination with agricultural activities is based on land use data obtained from the National Institute of Statistics (ISTAT), in particular the Agrarian Land Register of 1929 and the Agricultural Censuses of 1990 and 2020. While the timespan gives an interesting view on the major transformations during the last century, the categorisation between the datasets needs to be managed with attention, especially since the Agrarian Land Register had different methods for data collection in comparison with the more recent ones. Therefore, we analysed the individual data categories in detail to understand their content and consequently created a comparative table that allowed us to relate the past and current data between each other (Table 3.1).

Table 3.1 Comparative diagram of agricultural land use categories in the 1929 Agricultural Land Register and Agricultural Censuses of 1990 and 2020

Agricultural Land Register 1929		*Agricultural Census 1990–2020*	
Arable and forested land *(Superficie agraria e forestale)*	Arable land (simple and wooded plants) *(Seminativi [semplici e piante legnose])*	Utilised Agricultural Area *(Superficie Agricola Utilizzata)*	Arable land + family gardens *(Seminativi + orti familiari)*
	Meadows *(Prati e pascoli permanenti)*		Meadows *(Prati permanenti e pascoli)*
	Specialised woodcrops *(Colture legnose specializzate)*		Specialised woodcrops + vineyards *(Coltivazioni legnose agrarie + vite)*
	Woodland (fruit chestnut trees and other woodland) *(Boschi [castagneti da frutto e altri boschi])*	Total farm area *(Superficie Agricola Totale)*	Woodland related to farms + woodland for wood production related to farms *(Boschi annessi ad aziende agricole + arboricoltura da legno annessa ad aziende agricole)*
	Non-cultivated productions *(Incolti produttivi)*		Not used agricultural surface and other surfaces *(Superficie agricola non utilizzata e altra superficie)*

From the point of view of the categories of land use, the Land Register files (1929) indicated two different classes: 'arable and forested land' and 'unproductive' areas. The former was divided into five sub-classes: arable land, permanent meadows and pastures, specialised wood crops, woodland, non-cultivated productions. The distinction between specialised wood crops and woody plants is due to the large presence of the so-called *coltura promiscua* (mixed cultivation) which relates to the multiple use of the soil taking advantage of the different height development of plants (e.g., vine/olive and horticultural crops) in order to exploit, as much as possible, the few spaces available because of the complicated morphology that characterises the region (Quaini, 1973). In the modern agricultural ISTAT census campaigns, the categories are first divided into Total Farm Area (FA) and Utilised Agricultural Area (UAA), while in the 1929 Agrarian Land Register these are both included in the category 'arable and forested land'.

Ligurian Alps (Nervia and Argentina valleys)

The first case study area is located in the Ligurian Alps, Province of Imperia, inland the Western Riviera near the Italian–French border. The area today partly falls in the Alpi Liguri Regional Natural Park, instituted in 2007 to protect the Ligurian alpine ridge. The municipalities under analysis are entirely located within the Tyrrhenian catchment, which faces South and has a milder climate compared to the Po Valley catchment. However, the area is very diverse in terms of climatic and environmental factors as villages facing the coast show Mediterranean features while the highest sections are already in the Alps. The area is dominated by Monte Saccarello (2,200 m asl) in the municipality of Triora, the highest mountain of the region which is part of the main alpine ridge between Italy and France from which the Nervia and Argentina Valleys generate. The municipalities that suffered the greatest decrease of UAA in the ranges 1929–1990–2020 are: Triora from 3,497 ha to 2,485 ha and 223 ha in the upper Argentina Valley; Pigna from 3,117 ha to 1,454 ha and 79 ha in the upper Nervia Valley. Particularly interesting is the decrease between 1990 and 2020 (−91.03% and -94.57%, respectively). In this area, located in the alpine sector, the productive agricultural surface in some cases passed from more than 3000 ha to less than 300 ha between 1929 and 2020. Whilst generally associated with constant and often dramatic loss of agro-silvo-pastoral practices such remarkable decline is also given to the fact that forestry farms have been excluded by the ISTAT agricultural census from 2000 with a consequent decline of the overall surface (Liguria Ricerche, 2011).

Ligurian Apennines (Scrivia and Trebbia valleys)

The second case study is located in the inner part of the Genoese Apennines, beyond the main ridge which separates the Tyrrhenian from the Po Valley catchment in an area included in the Antola Regional Natural Park, instituted in 1995. Both the Scrivia and Trebbia Valleys are generated from the mountain system of Monte Antola (1,598 m asl), one of the highest peaks of the Ligurian Apennines. The Scrivia and Trebbia Rivers stream from Monte Antola northwards and join the Po respectively near Voghera (Lombardy) and Piacenza (Emilia-Romagna). The two valleys face North, and this is reflected in the climate, which is mainly continental at low elevation and of Apennine type at medium-high elevations. If, due to global warming, today the climate is considerably milder and drier, in the past the area was characterised by long, snowy and cold winters and short, cool summers with hardly any influence of the Mediterranean Sea. This is reflected on the type of agricultural production of municipalities of the Scrivia and Trebbia Valleys, which has been traditionally poor and very much conditioned by harsh environmental conditions. The comparison of the UAA through time shows a constant decline throughout the study area with some particularly meaningful cases. Some municipalities show a constant loss of productive agricultural surface in the time range 1929–1990–2020: Gorreto (Trebbia Valley) is the place with the sharpest decline as the UAA passed from 976 ha to 69 ha to 6 ha (−92.84%, −91.42%); Torriglia (Scrivia Valley) passed from 3,420 ha to 639 ha and 101 ha (−81.29%, −84.22%). Whilst being generally an issue throughout the area, the loss of UAA seems to have been more impactful in the highest and remotest villages of the Apennines and slightly less along valley floors.

The comparative analysis has been applied to both case studies at the municipality level, illustrating the extreme and opposite evolution of the dominant land uses (woodland and agriculturally used land) through time. These trends resulted in landscapes that completely changed spatial and existential character in only a few decennia (Dossche et al., 2016) with different challenges to tackle. Therefore, consequential shifts and even conflicts in views, decisions and strategies for future development are inevitable at every level of landscape management (from local farmers to national/European legislation).

Conservation policies and wilderness tourism: the role of private and public actors

At the institutional level, the Ligurian regional authority and other institutions (chiefly protected areas including Alpi Liguri and Antola Regional Natural Parks) have the complex role of protecting biodiversity linked to open areas according to national and European regulations, promoting the sustainable use of natural resources also to the light of new opportunities provided by rewilding (e.g., forestry) and reducing potential human-nature conflicts. In this sense, European Union projects (including Life projects such as WolfAlps EU, Interreg such as ALCOTRA projects like Biodiv'Alp and MARITTIMO like Castour/CamBioVIA) are crucial to reach these targets and to promote adaptation to a new environmental and climatic context both at regional and European level. The success of such initiatives is strictly tied to the promotion and encouragement of decision-making at the local level through the involvement of local stakeholders and bottom-up processes.

Alongside reforestation, another controversial consequence of rewilding is the return of animal species once disappeared or significantly reduced. From abandoned and rewilded rural areas, in recent, times these animals are increasingly seen in cities, a phenomenon which poses several challenges and risks, and has been controversially received by the public opinion (Pampaloni & Brocada, 2022). In the countryside, there is a degree of hostility towards animals that have an impact on rural activities (e.g., the wolf and the wild boar). Especially agricultural activities and livestock farms are threatened respectively by ungulates and wolves. Projects like WolfAlps EU promote peaceful coexistence between wolves and farmers through the adoption of specific measurements including the use of movable fencing combined with guard dogs.

Urban dwellers on the other hand have a different and possibly more positive perception on this matter, according to dynamics that are common to the rest of Europe (Smith, 2023). Even in this case, the perception of the phenomenon by public opinion is often driven by myths and fake beliefs spread by the web via social network and not adequately contrasted by effective communication campaigns (Nanni et al., 2020). The issues of rewilding and animal return in the Ligurian countryside remain something generally divisive and often a controversial point also in the political debate.[6] The increase of green areas is generally seen as something positive for the environment but also for the community and this is certainly true in most cases, but as explained above the situation is not always clear-cut.

There is a growing number of professionals who in recent times have pioneered the potential opportunity of nature-based economies in the context of rewilding through wildlife watching taking advantage of a renewed interest for wilderness tourism (Zanolin & Paül, 2022). Significant examples are the activities carried out by freelance photographers Paolo Rossi and Nicola Rebora in the Ligurian Alps and Apennines, the project *Wild horsewatching – I cavalli selvaggi dell'Aveto* coordinated by Evelina Isola and Paola Marinari of the association ReWild Liguria and by Nicolò Garaventa in the Antola Natural Park (Scrivia-Trebbia Valleys).

Using camera traps and direct observation, they all contribute to unveil to the public the secret aspects of Ligurian wildlife. Both Rossi and Isola acknowledge the importance of involving the local population whose approach to their activities is diverse, from open opposition to strict collaboration and acknowledgement of the importance of their work to revitalise the economy of such depressed areas.

Somehow paradoxically, however, the same rewilding processes that lead to the return of charismatic species are resulting in the contraction of clearings and open areas that are essential for wildlife watching, threatening their commercial success in Liguria. According to Rossi, wildlife tourism in Liguria is not comparable to other areas of the Alps and the Apennines (notably Gran Paradiso National Park and the parks of Abruzzo) where high-altitude grasslands allow the observation of many animals.

The need of open areas for the conservation of biodiversity linked to habitats considered as a priority by EU regulations, makes interventions by private and public institutions crucial to preserve such remnants of cultural landscapes. In addition to the conservation of priority environment and species, such efforts would also guarantee landscape openness and visibility, beneficial to nature-based economies linked to wilderness and green tourism similarly to what is happening elsewhere in Europe.[7]

Conclusions

As shown by the data discussed in this paper – notably with the two case studies presented above – the region of Liguria is at the frontline of rewilding processes not only in Italy but also in Europe, a situation which poses great challenges in terms of both management at local and European level and public opinion awareness. Although very preliminary and by their very nature limited to small groups, wildlife-watching tours provide evidence of the opportunity of nature-based economies in the Ligurian countryside, but also of human-animal coexistence. Other potential opportunities are given by the large availability of wood and timber for the promotion of forestry operations, an economic sector that is underdeveloped in the region. It is possible, however, that unregulated rewilding in the end might be an obstacle to the practical implementation of nature-based economies; this is particularly true of mountain areas where lack of landscape maintenance is directly associated with hydrogeological risk and the consequent significant limitations to accessibility and fruition.

In the context of the new strategy for biodiversity (2030) and carbon neutrality (2050) reforestation processes can help to reach one of the targets of the European Commission, which is the plantation of 3 billion trees by 2030 (European Commission, 2020). However, the implementation of such policies, particularly in Italian and Mediterranean mountains, must go through an analysis of gains and losses of such phenomena at the local level and above all there have to be specific management plans.

Therefore, considering on the one hand the rapid and extreme changes Mediterranean mountain landscapes underwent, and on the other hand the multi-layeredness of the landscape, its (local) stakeholders and (inter)national rules and regulations, it is fundamental to create context-depending solutions and strategies for future sustainable rewilding. Landscapes can have 'wilder' areas, but we should not forget the role of agriculture in a formerly rural context which produced cultural landscapes through a process of 'domestication' of nature which contributed to shaping a classic image of Liguria. A one-direction strategy towards reforestation and nature conservation is not possible, but rather a coexistence of Ligurian 'domestic' and 'rewilded' natures. Multifunctional landscapes where different assets and economies (nature-based or not) can be contemporary and complementary present would be the most sustainable outcome. More in general, a new paradigm through the encouragement of nature-based economies and local community-based decision-making processes (Dossche, 2022) appears as the most effective and realistic way to promote sustainable management practices in the Ligurian mountains which acknowledge both recent and somehow inevitable rewilding processes and the rich cultural and social legacy of the landscape.

Notes

1 For the purpose of this work, the term rewilding refers to a spontaneous process given by lack of management strategies, very typical of rural areas of Southern Europe. This is very different from the prevalent idea of rewilding, seen as a strategy which 'aims to restore self-sustaining and complex ecosystems, with interlinked ecological processes that promote and support one another while minimizing or gradually reducing human interventions' (Perino et al., 2019, p. 1).
2 Interviews were carried out between 2022 and 2024 and some of the authors participated to a guided excursion for wildlife watching by Paolo Rossi in the Trebbia Valley in June 2024. In addition to deep immersion in natural areas, such activities have also entailed informal interaction with local inhabitants on wildlife-related stories and more in general on the social and environmental aspects of rural areas of Liguria.
3 Data extracted from the 2019 Land Use Map by Regione Liguria.
4 An administrative division of Italy, operative since 2015, which consists in a special type of province for the regional capitals and some medium-large cities. This division often has some administrative contradictions mainly in terms of territory and landscapes included/excluded and of urban planning (cf. Brocada et al., 2022).
5 As an example, in 2022 the mayor of Genoa declared 'let's debunk the cliché that Genoa is a non-green city […] thanks to our mountains' (Canessa, 2022).
6 As shown, for example, by some 'anti-wolf' campaign posters by the political party Lega in the European elections of 2024.
7 Nature-based economies relate to business opportunities linked to rewilding that bring together conservation and entrepreneurship. It is a new economic paradigm carried out by associations such as Rewilding Europe (with practical examples including one in the Abruzzo region in Italy) and Rewilding Britain (cf: https://rewildingeurope.com/rewilding-in-action/nature-based-economies/; https://www.rewildingbritain.org.uk/why-rewild/benefits-of-rewilding/nature-based-economies).

References

Agnoletti, M., Piras, F., Venturi, M., & Santoro, A. (2022). Cultural values and forest dynamics: The Italian forests in the last 150 years. *Forest Ecology and Management*, *503*, Article 119655. https://doi.org/10.1016/j.foreco.2021.119655

Balzaretti, R., Pearce, M., & Watkins, C. (Eds.). (2004). *Ligurian landscapes. Studies in archaeology, geography & history*. Accordia Research Institute, University of London.

Brengola, A. (1939). La Carta forestale d'Italia. *La Rivista Forestale Italiana*, *1*, 7–10.

Brocada, L., Dossche, R., Mangano, S., Piana, P., & Priarone, E. (2023). Dinamiche di rinaturalizzazione tra criticità e opportunità di sviluppo: il caso della Liguria. In V. Albanese, & G. Muti (Eds.), *Oltre la Globalizzazione. Narrazioni/Narratives*, Vol. XXIII (pp. 791–797). Società di Studi Geografici.

Brocada, L., Mondino, L., & Priarone, E. (2022). La Città Metropolitana di Genova: analisi di aspetti geografici e criticità per un riordino territoriale. In F. Amato et al. (Eds.), *Catene/chains, Memorie geografiche*, Vol. XXI (pp. 925–932) Società di Studi Geografici.

Canessa, F. (2022, April 26). Sorpresa: "Genova è tra le città più verdi d'Italia". E il Comune vuole piantare ancora più alberi. *Genova24*. https://www.genova24.it/2022/04/sorpresa-genova-e-tra-le-citta-piu-verdi-ditalia-e-il-comune-vuole-piantare-ancora-piu-alberi-302220/
Cevasco, R. (2007). *Memoria Verde. Nuovi spazi per la geografia*. Diabasis.
Copernicus. (1990). *CORINE land cover 1990 (vector/raster 100 m), Europe, 6-yearly*. https://doi.org/10.2909/5c1f2e03-fcba-47b1-afeb-bc05a47bada0
Dossche, R. (2022). *Towards a sustainable rural mountain landscape: Exploring the (hi) stories of Val Borbera (Northern Apennines, Italy)*. Ghent University Press.
Dossche, R., Rogge, E., & Van Eetvelde, V. (2016). Detecting people's and landscape's identity in a changing mountain landscape. An example from the northern Apennines. *Landscape Research*, *41*(8), 934–949. https://doi.org/10.1080/01426397.2016.1187266
European Commission. (2020). *EU biodiversity strategy for 2030 – Bringing nature back into our lives*. https://environment.ec.europa.eu/strategy/biodiversity-strategy-2030_en
Faccini, F., Paliaga, G., Piana, P., Sacchini, A., & Watkins, C. (2016). The Bisagno stream catchment (Genoa, Italy) and its major floods: Geomorphic and land use variations in the last three centuries. *Geomorphology*, *273*, 14–27. https://doi.org/10.1016/j.geomorph.2016.07.037
Ferretti, F., Sboarina, C., Tattoni, C., Vitti, A., Zatelli, P., Geri, F., Pompei, E., & Ciolli, M. (2018). The 1936 Italian Kingdom Forest Map reviewed: A dataset for landscape and ecological research. *Annals of Silvicultural Research*, *42*(1), 3–19.
Gasparini, P., Di Cosmo, L., Floris, A., & De Laurentis, D. (2022). *Italian national forest inventory. Methods and results of the third survey*. Springer Nature.
Hearn, R., Watkins, C., & Balzaretti, R. (2014). The cultural and land use implications of the reappearance of the wild boar in North West Italy: A case study of the Val di Vara. *Journal of Rural Studies*, *36*, 52–63. https://doi.org/10.1016/j.jrurstud.2014.06.004
Il Secolo XIX. (2022, November 5). *Verde urbano, Genova è la città metropolitana con la più alta quota di territorio ricoperto di alberi (72%.)*. https://www.ilsecoloxix.it/liguria/2022/11/05/news/verde_urbano_genova_e_la_citta_metropolitana_con_la_piu_alta_quota_di_territorio_ricoperto_di_alberi_72_e_arrivano_20_m-12220979/
Liguria Ricerche. (2011). L'agricoltura in Liguria: rapporto sui dati provvisori del 6° Censimento generale dell'Agricoltura in Liguria. https://www.istat.it/it/files/2011/11/Rapporto_sui_dati_provvisori-.pdf
Moreno, D. (1990). *Dal documento al terreno. Storia e archeologia dei sistemi agro-silvo-pastorali*. Il Mulino.
Munafò, M. (Ed.). (2022). *Consumo di suolo, dinamiche territoriali e servizi ecosistemici*. Report SNPA 32/22. https://www.snpambiente.it/snpa/consumo-di-suolo-dinamiche-territoriali-e-servizi-ecosistemici-edizione-2022/
Nanni, V., Caprio, E., Bombieri, G., Schiaparelli, S., Chiorri, C., Mammola, S., Pedrini, P., & Penteriani, V. (2020). Social media and large carnivores: Sharing biased news on attacks on humans. *Frontiers in Ecology and Evolution*, *8*, Article 71. https://doi.org/10.3389/fevo.2020.00071
Pallotta, E., Boccia, L., Rossi, C. M., & Ripa, M. N. (2022). Forest dynamic in the Italian Apennines. *Applied Science*, *12*(5), Article 2474. https://doi.org/10.3390/app12052474
Pampaloni, C., & Brocada, L. (2022). Urban wildlife: l'inselvatichimento dello spazio urbano. In A. Primi & L. Brocada (Eds.), *Selve urbane: percorsi di ricerca* (pp. 17–30). GUP.
Perino, A. et al. (2019). Rewilding complex ecosystems. *Science*, *364*(6438), Article eaav5570. https://doi.org/10.1126/science.aav5570
Piana, P., Brocada, L., Hearn, R., & Mangano, S. (2024). Urban rewilding: Human-wildlife relations in Genoa, NW Italy. *Cities*, *144*, Article 104660. https://doi.org/10.1016/j.cities.2023.104660

Pincin, A. (2020). *La città rurale. Paesaggi in continuo divenire*. Asterios.
Quaini, M. (1973). *Per la storia del paesaggio agrario in Liguria*. CCIAA.
Regione Liguria. (2018). *Uso del Suolo sc. 1:10000 – Ed. 2018*.
Rota, M. P. (1991). La copertura vegetale della Liguria costiera. Dalla antropizzazione alla rinaturalizzazione. In A. Vallega (Ed.), *La Liguria e il mare* (pp. 125–146). Pubblicazioni dell'Istituto di Scienze Geografiche dell'Università di Genova, Facoltà di Magistero.
Smith, I. (2023, May 19). Rural vs urban divide: Why has the protection of wolves become so politicised in Europe? *Euronews.green*. https://www.euronews.com/green/2023/05/19/the-wolf-does-not-belong-here-why-has-a-conservation-success-become-a-heated-debate-in-eur
Tonini, M., D'Andrea, M., Biondi, G., Degli Esposti, S., Trucchia, A., & Fiorucci, P. (2020). A machine learning-based approach for wildfire susceptibility mapping. The case study of the Liguria region in Italy. *Geosciences*, *10*(3), Article 105. https://doi.org/10.3390/geosciences10030105
Zanolin, G., & Paül, V. (2022). Exploring the sustainability of wilderness narratives in Europe. Reflections from Val Grande National Park (Italy). *Geographical Review*, *112*(3), 444–465. https://doi.org/10.1080/00167428.2020.1869905

4 Nature in food and food in nature

Representing, inventing and practising naturalness and mountainness

Giacomo Pettenati

Introduction

As Goodman (1999) observes, the concept of nature has been largely depoliticised in the field of food studies, particularly in the area of agri-food studies. This depoliticisation has resulted in a perception of nature as a detached entity, separate from humans, which in turn has reinforced a dichotomy between nature and society. This dichotomy is a key focus of the social nature perspective, which seeks to challenge and reframe our understanding of the relationship between nature and society.

A critical and relational approach to food geography (Goodman, 2016; Murdoch, 2005) necessitates an alternative perspective on nature, one that is attentive to its social and political production and mobilisation in narratives and discourses that either support or oppose specific food production systems and food-place relations.

This chapter, based on the findings of an ongoing qualitative research project examining the politics of place in relation to the process of food heritagisation in the Italian Alps, focuses on the discursive mobilisation of the concept of nature within different narratives about what we may term "food mountainness", with a particular focus on Alpine cheese. As with the concept of nature, the definition of a mountain is a particularly fruitful area of investigation for post-structuralist and constructivist geography. This is true both in terms of the physical delineation of mountainous regions and the socio-anthropological characteristics attributed to mountainous societies (Debarbieux & Rudaz, 2015).

After an overview of the scholarly debate concerning the links between food and nature, the article turns to three distinct yet interrelated discourses on cheese naturalness: the discursive construction of natural cheese as a category, as espoused by Slow Food; the role of the presumed naturalness of some mountain cheeses in contributing to their quality; the vibrant debate about the coexistence in mountain regions between "natural" milk and cheese producers and the equally "natural" big carnivores threatening the animals they breed and graze.

DOI: 10.4324/9781003465164-6

Food and nature

Food is a primary medium for material and symbolic interactions between individuals, society and nature (Murdoch & Miele, 1999; Murdoch et al., 2000; Winter, 2005). Foods can be interpreted as liminal elements that connect nature and culture, the interior of the body and the external environment (Bell & Valentine, 1997). They can also be considered hybrid entities that combine and assemble organic and inorganic, material and symbolic, social and natural elements. The socio-natural essence of food is described by Alkon (2013, p. 675), who highlights the fact that it is 'one of the more obvious examples in which nature and human labour are inextricable'. Furthermore, given that food is currently the subject of vital social movements, it comprises an important standpoint from which to examine the politics of socio-nature on the ground.

The evolution of the intricate interrelationships between food and nature, and particularly the discourses constructed around them, can be encapsulated by focusing on two principal, seemingly contradictory processes: denaturalisation and renaturalisation.

The first process, de-naturalisation, is directly correlated with the evolution of human civilisation. This is evidenced by the advent of cooking techniques and the dissemination of agriculture, which represent pivotal stages in the conceptualisation of food as a cultural artefact, based on the (more-than) human transformation of natural raw materials (Levi-Strauss, 1964; Montanari, 2006). The process of 'liberating' food production from natural processes (Morgan et al., 2006) has been significantly accelerated by the advent of industrial society, which has both materially and ideologically 'domesticated' nature, placing it at the service of the principles of efficiency and profitability of the capitalist agro-industrial regime (Murdoch & Miele, 1999; Murdoch et al., 2000), as well as of the forced modernisation programmes implemented by state-led collectivist regimes (Deutsch, 2019).

The second process, re-naturalisation, is related to the role that nature plays in the social, cultural, and economic construction of food quality. Since the 1990s, the agri-food sector has been characterised by a shift towards quality (*quality turn*), largely driven by growing consumer mistrust in agro-industrial production systems (Goodman, 2003; Murdoch et al., 2000). In this context, the presumed naturalness of food has begun to constitute one of the key elements of the complex process of negotiated quality construction by the actors of the food system (Murdoch & Miele, 1999; Winter, 2005), together with other attributes such as provenance or tradition. In other words, the concept of nature has become a key element of the contemporary discourse surrounding the notion of 'good food', influencing consumer choices and shaping producers' strategies (Shugart, 2014). The new meanings attributed to the natural qualities of food, however defined, have led to the emergence of new consuming cultures and practices (Goodman & DuPuis, 2002). Sassatelli (2015) linked this to the general desire of consumers for a 'reconnection' with food production (Winter, 2005), the role of food choices in self-positioning and representing oneself within social relations (Bell & Valentine, 1997), and the production of 'ethical foodscapes', which are in opposition to the placeless foodscape of standardised industrial food (Morgan, 2010). As was to be expected, the market responded promptly to these new consumers' concerns (Murdoch et al., 2000) through the commodification of nature's representations, whereby naturalness is predominantly employed as a quality attribute of food, as perceived by consumers, and promoted by producers and distributors (Guthman, 1998).

Not considering here the actual contribution of natural elements in determining the organoleptic characteristics of food, it is possible to posit that the role of nature in the narratives about food quality is twofold. On the one hand, the purported 'naturalness' of the geographical contexts of production is invoked as a causal factor in the determination of food quality. The quality of places and the quality of food are thus socially and politically co-produced. Often, this mutual relationship is certified by geographical indication labels (Rippon, 2014), which associate the quality of the labelled agri-food products with that of their places of production. Such labelling schemes include those applied to food produced within the boundaries of protected areas, as exemplified by the Italian Gran Paradiso National Park (Beltramo et al., 2022). Similarly, there are noteworthy initiatives in the field of place/food branding that emphasise the interconnection between environmental quality and the "naturalness" and quality of local production. The 100% Valposchiavo project in Switzerland (Gros-Balthazard, 2022) is a case in point.

Conversely, there has been a notable increase in consumer interest in natural food products, which represents a significant aspect of the aforementioned 'quality turn'. The term "natural" is typically associated with food products that have undergone minimal human intervention during the transformation of raw materials into processed food items. In particular, consumers anticipate that natural food will also be organic, local, minimally processed, GMO-free, and free of artificial ingredients, antibiotics, hormones, and other substances that may impact animal and human health (Schirmacher et al., 2023). This generates a broad and diverse semantic cloud where the meaning of nature is continuously produced and reproduced.

The concept of natural food is not without its political implications, serving as a symbol of alternative food chains that challenge the conventional, globalised agri-food system. Two of the principal "natural food movements" pertain to wine and cheese production. The limited human control over the processes of fermentation that generate these products from raw materials is a politically charged issue, representing a reaction against the market-oriented approach and the normativity of conventionally labelled products (e.g., PDO wines and cheeses) (Pavoni, 2020).

The concept of organic food represents an intriguing nexus between two distinct narratives surrounding the naturalness of food. On the one hand, there is the notion of food as a product of a process involving the transformation of raw materials. On the other hand, there is the perspective that the quality of food is contingent upon the environmental resources that combine to form its production sites. A central tenet of the 'organic conceptual narrative' (Lockie et al., 2002) is the interconnection between the natural characteristics associated with food products and the environmental attributes of the geographical areas where food is produced (Guthman, 2014; Ilbery & Kneafsey, 2000).

The mutual relations between food and nature are also the object of gastropolitical narratives (Appadurai, 1981), which support biopolitical (Goodman et al., 2017) and moral economy (Jackson et al., 2009) discourses concerning the presumed higher value of some food production systems, as more representative of specific social or cultural groups. Furthermore, the notion of an ostensible detachment from natural processes is a recurring theme in the discourses espoused by movements that are politically opposed to certain food industrial innovations, such as cultivated meat, which has only recently emerged as a topic of contention (Bertero et al., 2023).

One area in which the politicisation of nature is particularly evident is in the discourse surrounding cheese production and consumption. This is discussed in the following section, which focuses on three key debates and practices within this field.

Discourses around cheese naturalness

Cheese exists in many food cultures since ancient times, serving as a means of preserving milk for extended periods and facilitating the trade of dairy products (Kindstedt, 2012). The production of cheese is the result of a complex combination of material and immaterial elements, derived from the coagulation of casein triggered by rennet, which occurs during the transformation of milk from different domesticated animals. The manner in which these elements (breeding methods; transformation techniques, cultural factors, pasture characteristics, conservation methods, and so forth) are integrated is frequently the subject of political processes, including negotiations and conflicts surrounding the utilisation of resources.

In the last two decades, numerous contributions have addressed various geographical issues related to cheese. These include the role of cheese production in shaping rural landscapes, societies and economies, and vice versa (Grosinger et al., 2022; Lamarque and Lambin, 2015; Tregear, 2003). The controversial process of recognising some dairy products as typical of certain geographical areas and the role of transcalar forces such as the market and industrialisation in shaping their characteristics (Grasseni, 2016) is also a topic of interest. Similarly, the possible perverse effects, especially environmental, of the economic success of some products (García-Hernández et al., 2022). Furthermore, the role of cheese as a cultural artefact in generating and transmitting territorial narratives has been elucidated by Berno and Fusté-Forné (2020).

Cheese production is far from being a 'natural' process, given the unavoidable human intervention involved in the addition of rennet to milk and the guidance of the cheesemaking process. Nevertheless, the role of nature in enhancing cheese quality and the significance of the interconnections between cheese production systems and the environmental context in which they are situated are prominent themes in the majority of discourses pertaining to diverse cheeses, particularly when they emphasise their geographical origins and their integration into territorial systems. Furthermore, the processes of industrialisation and the implementation of food safety regulations have resulted in a significant detachment between cheese and its terroir of production, particularly through the introduction of milk pasteurisation. Consequently, common statements like "every raw-milk cheese is an artefact of the land; it carries the imprint of the earth from which it came. A cheese…is never just a thing to put in your mouth. It's a living piece of geography" (Kessler, 2010, p. 58) only make sense when they refer to raw milk (like Kessler correctly does).

The mainstream narrative of cheese as a transformed terroir (Whited, 2018) is at odds with the actual production of most cheeses, which substantially deterritorialises this type of product. This is particularly evident in the material sense of disconnecting them from the influence of the place where they are produced, for example, through the killing of most bacteria that are naturally present in milk through pasteurisation.

Conversely, the inextricable link between cheese and its terroir (typically represented as nature) is a central theme in most discourses and strategies that seek to reterritorialise cheese. This often involves the production and practice of alternatives to the conventional agro-industrial food system and the standardisation of food culture.

In this sense, some cheese is considered by some consumers and some relevant actors of the food debate (i.e., Slow Food) as a "moral good". This designation situates consumers as "co-producers", involved in acts of moral salvation and ecological responsibility by contributing to the protection of "traditional" lifeways and landscapes (MacDonald, 2013, p. 93).

The concept of nature is invoked in discourses pertaining to cheese, encompassing notions of local embeddedness, authenticity and the political implications of its reterritorialisation. This paper focuses on three illustrative examples, based on empirical research conducted in the Italian Alpine context.

Slow Food's natural cheese campaign

Although the natural wines movement is widely recognised, despite the fact that its very definition is open to question, the parallel movement in cheese is much less well known, with the exception of a small number of producers, consumers and food activists. One of the principal actors in the production and dissemination of narratives and definitions pertaining to the concept of "natural cheese" is the Slow Food movement, which also plays a pivotal role in the formation and sustenance of a global network of individuals and organisations that espouse and facilitate the production of such cheese. In accordance with the Slow Food definition, cheese can be classified as natural if it meets two fundamental criteria: (a) the product is manufactured from raw (i.e., non-pasteurised) milk; and (b) no selected industrial starters are employed to initiate the process of fermentation, whereby milk is transformed into cheese. These two criteria, which may appear to be process technicalities, are in fact highly political, as they are in direct opposition to the prevailing regulations, beliefs, and practices in terms of: (a) food safety and (b) industrial control of the cheesemaking process. The utilisation of raw milk is discouraged, or even prohibited, in numerous European countries due to the tangible risks of human contamination with specific bacteria that may be present in milk. Nevertheless, its proponents maintain that it is the sole means of ensuring that the terroir of the production area exerts an actual influence on the final product. This is due to the bacterial flora that is naturally present in milk, the composition of which is directly related to the grazing practices of animals and the ecosystems of pastures. However, this flora is essentially destroyed through pasteurisation. The utilisation of pasteurised milk, which has been devoid of the natural starters that are integral to the cheesemaking process, inevitably necessitates the incorporation of added starters. In the majority of cases, these are selected on an industrial scale and lack any connection with the provenance of the cheese in question.

In Slow Food's narratives and "naturalist aesthetics" (MacDonald, 2013, p. 98), the significance of natural cheese extends beyond the technical process of transforming milk into cheese. It is also closely associated with broader concerns, such as animal welfare, human health and sustainable resource management. This illustrates the political and territorial dimensions of their campaigns and advocacy activities in support of natural cheese:

> The term "natural" is used to describe artisanal cheeses that are produced without the use of industrial starters. This aspect represents merely the initial stage of an alternative model to intensive and standardised production. The production of cheese without the use of industrial starters necessitates the implementation of a well-managed farming system, characterised by the presence of healthy animals that are permitted to graze for a significant portion of the year. Additionally, the process does not entail the pasteurisation of milk, a practice that is unnecessary in this context. This approach ensures the preservation of the aromatic integrity and nutritional value of raw milk. Furthermore, it requires the expertise of a skilled cheesemaker, capable of producing cheese without the use of shortcuts or corrective measures. In conclusion, natural cheese is the outcome of an alternative production philosophy, the sole one capable of offering superior quality cheeses, while respecting the environment and animal welfare.[1]

Despite the influence of consumer preferences, which value the social status and identity associated with the consumption of natural products (MacDonald, 2013, p. 98), Slow Foods explicitly positions its campaign in support of natural cheese as a highly political endeavour, offering an alternative to the dominant agrifood system. The organisation bestows a prize upon cheesemakers who distinguish themselves for practising the natural cheese philosophy, which is known as the "Resistenza casearia", which could be roughly translated as "dairy resistance".

Nature in cheese. The influence of the terroir on cheese quality

The environmental characteristics of the places of production play a pivotal role in determining food organoleptic qualities, as evidenced in discourses on food typicality and geographical indication regulations. These characteristics represent a fundamental aspect of food quality, alongside historical rootedness in local food culture. The concept of terroir, initially developed to describe the link between wine characteristics and geology, has recently become an umbrella term, more generally referring to the influence of the geographical and cultural characteristics of the area of production on food products.

In the context of cheese, the concept of terroir is frequently employed in a rural determinist framework, whereby the direct influence of pasture vegetation, climate and geology on cheese taste and organoleptic characteristics is emphasised. The term "rural determinism" is used to describe the belief that plants and animals, as well as the products derived from them, are characterised by the conditions of the geographical place where they have grown, been raised or produced. This implies that the geographical region, with its distinctive biophysical characteristics, cultural practices, traditions, and ways of thinking, exerts an influence on agricultural products and foodstuffs originating from that region (Nygård & Wramner, 2014, p. 336).

The relationship between cheese quality and the environmental characteristics of pastures is a central issue in many of the most studied conflicts related to geographical indication labels. This is particularly evident in the differentiation between cheese produced during the summer in high mountain pastures (in Italian, alpeggi) and cheese produced year-round, with animals also fed hay or fodder.

Among the most studied Italian cases is that of Bitto, a mixed milk cheese (cow-goat) historically produced in a limited number of alpine pastures of the Orobic Alps (Lombardy). The product's designation as a PDO (Protected Designation of Origin) in 1995 was accompanied by a significant redefinition of its production regulations, which resulted in the marginalisation of the practice of grazing animals in mountain pastures. This was done in order to expand the area and production period, with the aim of increasing the product's market presence. A network of producers based in the historical production area challenged the recommendations set forth in the PDO specifications, contesting the limited role attributed to the high mountain pastures environment. In response, they established an alternative geographical indication scheme (Slow Food Presidia), reserved for a cheese they have designated "Storico Ribelle" (Historic Rebel).

The narratives constructed around the distinctive qualities of Storico Ribelle place emphasis on the intrinsic connection between the natural characteristics of the pastures where it is produced, the grazing and cheese-making techniques that have historically been developed on those pastures due to their environmental attributes, and the quality of the cheese:

> The livestock eat only pasture grass. The regulations prohibit the use of feeds and silage. In the Orobic Alps the plentiful fresh water, sunshine and differences in altitude contribute to the presence of a rich floral variety, which in turn gives the milk its superior nutritional quality. The biodiversity of the mountain environment, it's perfumes and aromas, give unique characteristics to each wheel of Heritage Bitto produced in the few alpine locations of the Consortium.[2]

Nevertheless, what happened could also be interpreted as an attempt by some producers to secure their market niche, as argued by Rinallo and Pitardi (2019).

Nature versus cheese: sharing space with large carnivores

A third relevant area of debate about the discursive use of nature in cheese production is radically different from the two discussed above, where the presumed harmony between cheesemaking and nature is presented as essential to guarantee cheese quality. I refer to the highly conflictual debate about the impact of the increasing presence of large carnivores (especially wolves and bears) in the Italian Alps on the extensive grazing of high mountain pastures. According to recent surveys, around 1000 wolves live in the Italian Alps, whereas in the 1970s they were thought to be almost extinct.[3] While the rapid growth of the Alpine wolf population can be seen as a successful outcome of EU environmental projects (in particular LifeWolfAlps), its impact on human activities in mountain areas is highly controversial (De Biaggi & Walter, 2023). Historically, the coexistence between humans and wolves has always been archetypical of the separation between human and "beastly" spaces (Philo & Wilbert, 2000). Currently, in the Italian Alps, the most radical positions claiming the need for a neat separation between human and wolf "territories" come from actors involved in extensive grazing and often cheese-making. The work of shepherds and pastoralists has been objectively changed by the return of these large carnivores, whose memory was almost lost in the minds of humans and other animals (e.g., shepherd dogs), with the need to protect and guard grazing animals to avoid predation. Thus, paradoxically, the most controversial narratives about the relationship between nature and cheesemaking come from the same social group that produces and disseminates discourses about the naturalness and quality of cheeses produced in highly natural geographical contexts. This discursive division between a "good" nature, which favours human activities and the production of quality food, and a "bad" nature, which should be kept away from human territories through defensive strategies, clearly shows the social and political construction of this concept, also in its use in relation to food and, in particular, cheese.

As I will argue in the last section, this political use of nature in the mountain extensive grazing and cheese-making sector can be interpreted as an element of a broader process of discursively supported specific politics of place, which in the case of the mountain contexts discussed in this chapter can be defined as a "politics of mountainness".

Nature, food and the politics of mountainness

Nature, and the politics of nature related to the management of environmental resources and their relationship with human settlements and society, play a key role in this process of social and political construction of mountains as a category of thought and as a referent for collective action. For example, natural-claimed food is at the heart of the self-narratives circulated by chefs who position themselves as deeply rooted in mountain food culture. This is the case of the South Tyrolean three-star chef Norbert Niederkofler, who describes his cooking project "Cook the Mountain" as follows: "Nature is my natural dispensation, the best I could have. It is based on what nature offers me in a certain period of the year that I start to create my menu, not vice versa"[4] (Niederkofler, 2020).

"Harmony with nature" as a key element of the Alpine identity, foodscape and way of life is also highlighted in the Alpine Food Charter, a participatory document signed by more than 2000 "people who care about the future of the Alps2, which advocates the protection of the Alpine food heritage and supports its nomination to the UNESCO list of Intangible Cultural Heritage (Lapiccirella Zingari, 2019).

Conversely, references to nature are basically absent from the regulatory framework of the EU "Mountain Product" labelling scheme, which, at least in its Italian implementation, is assigned exclusively according to administrative and statistical criteria. Even if the role of natural factors in contributing to food quality is explicitly mentioned in the EU regulation that legally regulates the use of the label (No. 1151/2012), it can be applied to products whose production takes place entirely or, in some cases, partially in municipalities that are institutionally classified as mountainous, regardless of the specific characteristics of individual localities and production systems.

Nature is mobilised in the co-construction of quality between food and mountain localities. On the one hand, mountain products are a priori considered by consumers to be of higher quality, also because they are more "natural" (Oliveira Linder, 2021), regardless of the actual relationships between food production and environmental components (Varotto, 2020). On the other hand, narratives about the environmental value of mountain places and their attractiveness often include their role as places of production of presumably sustainable and "natural" food (Alderighi et al., 2016).

As the three examples presented above show, cheesemaking in mountain regions is a highly politicised arena where food biopolitics, which affect the complex network of more than human agents that contribute to cheese production (Clay & Yurco, 2020), combine with gastropolitical strategies (Appadurai, 1981) that aim to use the narratives about cheese and its relationship with 'nature' to support specific socionatures and politics of mountainness. These include narratives about the alleged bidirectional relationship between the quality of food (in this case cheese) and the quality of places, where the role of a variously defined 'nature' is widely exploited by territorial and food marketing.

Notes

1 https://www.slowfood.it/formaggi-naturali/ (Translation from the author; Last access: 17/03/2024).
2 https://storicoribelle.com/en/storico-cheese-production/ (Last access: 8/10/2024).
3 https://www.isprambiente.gov.it/it/attivita/biodiversita/monitoraggio-nazionale-del-lupo/risultati (Last access: 8/10/2024).
4 https://norbertniederkofler.com/en/i-m-cook-the-mountain (Last access: 8/10/2024).

References

Alderighi, M., Bianchi, C., & Lorenzini, E. (2016). The impact of local food specialities on the decision to (re) visit a tourist destination: Market-expanding or business-stealing? *Tourism Management*, *57*, 323–333. https://doi.org/10.1016/j.tourman.2016.06.016

Alkon, A. H. (2013). The socionature of local organic food. *Antipode*, *45*(3), 663–680. https://doi.org/10.1111/j.1467-8330.2012.01056.x

Appadurai, A. (1981). Gastro-politics in Hindu South Asia. *American Ethnologist*, *8*(3), 494–511.

Bell, D., & Valentine, G. (1997). *Consuming geographies. We are where we eat*. Routledge

Beltramo, R., Duglio, S., & Cappelletti, G. M. (2022). Should I stay or can I go? Accessible tourism and mountain huts in Gran Paradiso National Park. *Sustainability*, *14*(5), Article 2936. https://doi.org/10.3390/su14052936

Berno, T., & Fusté-Forné, F. (2020). Imaginaries of cheese: Revisiting narratives of local produce in the contemporary world. *Annals of Leisure Research*, *23*(5), 608–626. https://doi.org/10.1080/11745398.2019.1603113

Bertero, A., Biressi, S., Buscemi, F., Conti, L., Cresti, M., Gargioli, C., Lo Sapio, L., Loera, B., Poncibò, C., & Stano, S. (2023). We need an informed discussion on cultivated meat. Technical, social, and humanities specialists call for measured discussion on the future of food. *Nature Italy*. https://www.nature.com/articles/d43978-023-00056-1

Clay, N., & Yurco, K. (2020). Political ecology of milk: Contested futures of a lively food. *Geography Compass*, *14*(8), Article e12497. https://doi.org/10.1111/gec3.12497

De Biaggi, M., & Walter, T. (2023). Wolf–Human coexistence in the Alps: The LIFE WolfAlps EU project. *eco. mont-Journal on Protected Mountain Areas Research and Management*, *15*. https://doi.org/10.1553/eco.mont-15-1s46

Debarbieux, B., & Rudaz, G. (2015). *The Mountain*. The University of Chicago Press.

Deutsch, R. (2019). *The food revolution in the Soviet Union and Eastern Europe*. Routledge.

García-Hernández, C., Ruiz-Fernández, J., & Rodríguez-Gutiérrez, F. (2022). Geographical indications in cheese mountain areas: Opportunity or threat to landscape and environmental conservation? The case of Cabrales (Spain). *Applied Geography*, *146*, Article 102753. https://doi.org/10.1016/j.apgeog.2022.102753

Goodman, D. (1999). Agro-food studies in the 'age of ecology': Nature, corporeality, biopolitics. *Sociologia Ruralis*, *39*(1), 17–38. https://doi.org/10.1111/1467-9523.00091

Goodman, D., (2003). The 'quality turn' and alternative food practices: Reflections and agenda. *Journal of Rural Studies*, *1*(19), 1–7. https://doi.org/10.1016/S0743-0167(02)00043-8

Goodman, D., & DuPuis, E. M. (2002). Knowing food and growing food: Beyond the production–consumption debate in the sociology of agriculture. *Sociologia Ruralis*, *42*(1), 5–22. https://doi.org/10.1111/1467-9523.00199

Goodman, M. K. (2016). Food geographies I: Relational foodscapes and the busy-ness of being more-than-food. *Progress in Human Geography*, *40*(2), 257–266. https://doi.org/10.1177/0309132515570192

Goodman, M. K., Johnston, J., & Cairns, K. (2017). Food, media and space: The mediated biopolitics of eating. *Geoforum*, *84*, 161–168.

Grasseni, C. (2016). *The heritage arena: Reinventing cheese in the Italian Alps*. Berghahn Books

Gros-Balthazard, M. (2022). Émergence du fait alimentaire dans le Valposchiavo (Suisse) à travers la définition d'un projet de territoire «en transition». *Journal of Alpine Research| Revue de géographie alpine*, Article 110–112. https://doi.org/10.4000/rga.10889

Grosinger, J., Grigulis, K., Elleaume, N., Buclet, N., & Lavorel, S. (2022). Community-based institutions shape cheese co-production in a French Alpine valley. *Mountain Research and Development*, *42*(3), R25–R34.

Guthman, J. (1998). Regulating meaning, appropriating nature: The codification of California organic agriculture. *Antipode*, *30*(2), 135–154. https://doi.org/10.1111/1467-8330.00071

Guthman, J. (2014). *Agrarian dreams: The paradox of organic farming in California*. University of California Press.

Ilbery, B., & Kneafsey, M. (2000). Producer constructions of quality in regional speciality food production: A case study from southwest England. *Journal of Rural Studies*, *16*(2), 217–230. https://doi.org/10.1016/S0743-0167(99)00041-8

Jackson, P., Ward, N., & Russell, P. (2009). Moral economies of food and geographies of responsibility. *Transactions of the Institute of British Geographers*, *34*(1), 12–24.

Kessler, B. (2010). *Goat song: A seasonal life, a short history of herding, and the art of making cheese*. Scribner

Kindstedt, P. (2012). *Cheese and culture: A history of cheese and its place in western civilization*. Chelsea Green Publishing.

Lamarque, P., & Lambin, E. F. (2015). The effectiveness of marked-based instruments to foster the conservation of extensive land use: The case of Geographical Indications in the French Alps. *Land Use Policy*, *42*, 706–717. https://doi.org/10.1016/j.landusepol.2014.10.009

Lapiccirella Zingari, V. (2019). Alpine communities and their food heritage as intangible cultural heritage. Transnational participatory fieldwork. In E. Falk, & S. Park (Eds.), *Traditional food. Sharing experiences from the field* (pp. 36–54). ICHCAP.

Levi-Strauss, C. (1964). *Le Cru et le Cuit. Mythologiques, 1*. Plon.

Lockie, S., Lyons, K., Lawrence, G., & Mummery, K. (2002). Eating 'green': Motivations behind organic food consumption in Australia. *Sociologia Ruralis*, *42*(1), 23–40. https://doi.org/10.1111/1467-9523.00200

MacDonald, K. I. (2013). The morality of cheese: A paradox of defensive localism in a transnational cultural economy. *Geoforum*, *44*, 93–102. https://doi.org/10.1016/j.geoforum.2012.03.011

Montanari, M. (2006). *Food is culture*. Columbia University Press.

Morgan, K. (2010). Local and green, global and fair: The ethical foodscape and the politics of care. *Environment and Planning A*, *42*(8), 1852–1867. https://doi.org/10.1068/a42364

Morgan, K., Marsden, T., & Murdoch, J. (2006). *Worlds of Food*. Oxford University Press.

Murdoch, J. (2005). *Post-structuralist geography: A guide to relational space*. Sage.
Murdoch, J., Marsden, T., & Banks, J. (2000). Quality, nature, and embeddedness: Some theoretical considerations in the context of the food sector. *Economic Geography*, *76*(2), 107–125.
Murdoch, J., & Miele, M. (1999). 'Back to nature': Changing 'worlds of production' in the food sector. *Sociologia ruralis*, *39*(4), 465–483. https://doi.org/10.1111/1467-9523.00119
Niederkofler, N. (2020) *Cook the mountain. The nature around you*. Prestel.
Nygård, O., & Wramner, P. (2014). Terroir – A key concept when building a geographic identity for food. In P. Rytkönen (Ed.), *Food and Rurality in Europe*. COMREC. https://doi.org/10.1016/B978-0-444-52512-3.00157-1
Oliveira Linder, M. (2021). *Do mountains matter?: Italian consumer preferences regarding mountain food products and the mountain labelling scheme*. Doctoral Thesis in Mountain Environment and Agriculture, Free University of Bozen-Bolzano.
Pavoni, A. (2020). Profanating Gastro-Normativity: Exploring the Nonhuman Materiality of Making and Tasting Wine in the Natural Wine Movement. *Cultural Politics*, *16*(3), 367–386.
Philo, C., & Wilbert, C. (2000) *Animal spaces, beastly places*. Routledge.
Rinallo, D., & Pitardi, V. (2019). Open conflict as differentiation strategy in geographical indications: The Bitto Rebels case. *British Food Journal*, *121*(12), 3102–3118. https://doi.org/10.1108/BFJ-11-2018-0738
Rippon, M. J. (2014). What is the geography of geographical indications? Place, production methods and protected food names. *Area*, *46*(2), 154–162.
Sassatelli, R. (2015). Consumer culture, sustainability and a new vision of consumer sovereignty. *Sociologia Ruralis*, *55*(4), 483–496. https://doi.org/10.1111/soru.12081
Schirmacher, H., Elshiewy, O., & Boztug, Y. (2023). That's not natural! Consumer response to disconfirmed expectations about 'Natural' food. *Appetite*, *180*, Article 106270. https://doi.org/10.1016/j.appet.2022.106270
Shugart, H. A. (2014). Food fixations: Reconfiguring class in contemporary US food discourse. *Food, Culture & Society*, *17*(2), 261–281. https://doi.org/10.2752/175174414X13871910531665
Tregear, A. (2003). From Stilton to Vimto: Using food history to re-think typical products in rural development. *Sociologia ruralis*, *43*(2), 91–107. https://doi.org/10.1111/1467-9523.00233
Varotto M. (2020). *Montagne di mezzo*. Einaudi.
Whited, T. L. (2018). Terroir transformed: Cheese and pastoralism in the western French Pyrenees. *Environmental History*, *23*(4). https://doi.org/10.1093/envhis/emy057
Winter, M. (2005). Geographies of food: Agro-food geographies-food, nature, farmers and agency. *Progress in Human Geography*, *29*(5), 609–617. https://doi.org/10.1191/0309132505ph571pr

5 Authenticity (re-)naturalised

The mise-en-scene of raw meat in the heritagized city of Florence

Panos Bourlessas and Matteo Puttilli

Introduction

In 2018, Gravari-Barbas (2018) very characteristically described contemporary tourism as a "heritage-making machine", to highlight the parallel symbolic and physical heritage-related transformations (be them of sites, of objects, or both) induced by the phenomenon globally. As if in a variegated process of "accumulation by conservation" (Büscher & Fletcher, 2014), this machine seeks to produce and stabilize new heritage niches so as to guarantee the continuation of the tourist experience and to extend the respective margins for profit generation, circulation and accumulation. In this chapter, we depart from Gravari-Barbas' notion in order to reach some empirically-based reflections on a case study so emblematic to processes of urban heritagization overall, namely, Florence, Italy.

Lately, a peculiar phenomenon is noticeably transforming the city's heritagized retailscape: through an increasingly standardized practice of an interior-exterior design—or *exteriorizing* interior design—based on transparency, pieces of raw meat are being exposed behind the windows of Florentine restaurants, the latter rearranging their visual appeal by centring it on raw meat, and connecting to public space and gaze through this meat (see Figure 5.1). Following Castree's (2003) conceptualization of commodified nature, we approach raw meat as an "external nature" (Castree, 2003, p. 284), and more precisely as "externa*lized* nature" involved in commodification processes, the latter relating to wider phenomena of urban touristification and heritagization of our case study. Precisely, by externalized we enhance Castree's argument in order to suggest that, being (re)presented as raw, as unprocessed, the exposed meat performs a "nature" that is *made* external to the consumers, to consumption practices, and to urban space: creating a direct relation to what can commonly be perceived as "natural", the exposed meat conceals the entire production process in which the meat is already inserted even while "waiting for" its own consumption in the restaurants' windows.

In its spectacular outwardness, this recent design trend arguably transforms restaurants into boutique-like food spatialities, gradually converting the city's street-level visuality overall: it adds literal non-human flesh to the already

DOI: 10.4324/9781003465164-7

Figure 5.1 A tourist taking pictures of the raw meat exposed in the windows of one of the most iconic *bistecca* restaurants of the UNESCO area. Authors' picture.

aestheticized and spectacularized urban space. Indeed, the "tourist gaze" (Urry, 2002) cannot but be struck by the proliferating visibility of the exposed raw pieces of the *bistecca alla fiorentina* (or just *bistecca* or *fiorentina*), awaiting to get ingested in order to finally materialize the so much searched for "authentic" local experience, an authenticity that is reinforced by heritagization processes, whether already established or still ongoing. Ingestion is now mediated by the glass window; mediation completes the experience, it guarantees it; and visibility extends towards more-than-visual and atmospheric directions of a consumed authenticity (Zukin, 2011).

In this chapter, we focus on this transforming and equally transformative urban visuality of exposed raw meat to showcase how nature, and its selective instrumentalization, becomes a new "silent" key to the tourism-heritage-authenticity nexus unfolding throughout central Florence. We do so by focusing on how meat as exposed externalized nature impacts on the tourists' perceptions. Specifically, we aim to respond to the following questions: How does "nature" enter the consumptive entanglement between heritage, tourism and authenticity? In what ways does it contribute to the visitor's experience of eating the *bistecca* in specifically orchestrated culinary atmospheres? Finally, what is the contribution of "nature" in the heritage-making process? In terms of methodology, we address the above questions through a web-based critical discourse analysis of Bistecca-focused reviews of restaurants located in Florence's UNESCO area; moreover, a semi-structured interview with the Head of the *Accademia della Fiorentina* has been conducted. Our analysis illustrates that the staging of *bistecca* pieces is not limited to its visuality but, instead, this visuality extends outwards in order to construct culinary atmospheres where, by being immersed inside of it, the consumers construct the searched sense of local authenticity, thus satisfying their experiential demands.

Furthermore, and regarding the wider contribution of this text, we suggest that the phenomenon in question not only is exemplary to the dynamics between spaces, tastes and affect (Falconer, 2021), but has the potential to reveal (some of) these dynamics especially in relation to heritage and its fragile social construction. Precisely, by positioning it in relation to tourism as a heritage-making machine, we approach the raw meat's spectacular staging both as our analytical gate to the city's wider heritage-induced transformations and, most importantly to the purposes of this volume, as a local practice that performs the tourism-heritage-authenticity nexus throughout Florence's central urban space by foregrounding the instrumental yet implicit role of "nature" in advancing the nexus so as to hold its elements together. For the heritage-making machine does not create new heritage sites and objects alone without also creating new entanglements of already existing nexuses of commodification, which "nature" can be implicitly key to.

Food and heritage in urban space: the bistecca's hidden nature?

The last years have seen an intensification of the debates, both academic and public, that problematize the continuously increasing touristification of the city centre of Florence, together with its complex and contested socio-spatial transformations that affect also peri-central neighbourhoods. The city's UNESCO heritage area has been so dramatically affected by overtourism that has become an emblematic negative case that called for measures that can restrict the phenomenon's negative consequences, especially after the COVID-19 shock and the excessive recovery (Liberatore et al., 2022; Amore et al., 2020). These consequences arguably span across two major spheres of city life that are subject to transformations strongly sensed by city dwellers.

On the one hand, the rise of short-term and platform-based rentals (i.e., Airbnb) results in an increasing moving of more traditional and long-term residents out of the central areas and, therefore, in a significant change of the city's residential tissue overall, the latter characterized by a notable penetration of tourists in previously dominantly residential areas (Celata & Romano, 2020). On the other hand, the Florentine historical retailscape is undergoing the last years a proliferation of economic activities related to the consumption of foods and drinks, a process that has been termed as "foodification" (Loda, Bonati & Puttilli, 2020; Puttilli & Loda, 2022). It is the intersection of the two that is central to the phenomenon analysed here, as the foodification dynamics generated in the city are quite specific and strongly related to the increasing presence of short-term visitors, especially tourists, in the heritagized city centre. What is specific about the ongoing foodification dynamics is the character that arguably dominates the emergence of new spatialities for food consumption, as well as new visualities. These spatialities and visualities seek to increasingly target an international group of consumers that is often in search of authentic local experiences, the latter to be located in a city centre that is "fixed" and labeled as authentic through heritagization and its unfolding retailscape.

As already illustrated by one of the authors (Puttilli et al., 2020), the visibility performed by the exposition of the *bistecca* in the windows of restaurants is highly entangled with discourses and practices of cultural heritage on the one hand, and with the proliferating touristification of central Florence on the other. Indeed, in this framework it can be argued that the exhibition and increased visibility of the *bistecca* can work as an illuminating example of the paradoxes inherent to the processes of constructing and safeguarding heritage: on the one hand, the need to protect things, practices, places; on the other hand, the gradual and accumulative alienation of these elements from those discursively entitled as "typical" possessors, namely, the members of the local community. Working in concert with the official heritagization discourses and efforts, although unsuccessful for the moment as the 2018 candidacy of the *Bistecca alla fiorentina* as UNESCO cultural heritage did not reach a successful conclusion, the exhibition of the *bistecca*, we can argue, performs a well-curated visibility of the product as heritage-in-the-making, while, at the same time, further alienates it from the local population by fetishizing it and distancing from ordinary, local consumers and their practices.

In this chapter we take the above argument further in order to make space for "nature" as key—yet relatively *silent*—element of the evolving processes in question. By making space for "nature", we argue, it is possible to further deconstruct the "heritage-making machine" (Gravari-Barbas, 2018) revealing the complexity of this machine as well as, equally, (one of) the paradoxical relationships that make it possible. Indeed, after the massive displacement of slaughterhouses outside of urban centres in the West so that they become invisible to the public eye (Philo & MacLachlan, 2018), it seems that the heritage-making machine brought about a new visibility in Florence, this time carefully stylized and selective, in a period of progressive decrease of more traditional raw meat visibilities, namely those of old butcheries.

So, in what ways is "nature" present albeit silent in the *bistecca*? We suggest that these ways are two, which echo social constructivist approaches to nature (Mariyani-Squire, 1999). The first one relates to the geographical association of the product with extra-urban space. A common discourse and practice of locals is that the "right place" where to eat the bistecca is rural, rather than urban, space; in rustic environments close to nature. This association, never actually revealed in the touristic promotional materials relating to the product, is further reinforced by the fact that such peri-urban locations, such as the Mugello mountainous area, are home to the livestock most typical to the bistecca, that is, the Chianina breed. Like this, peri-urban nature becomes the element and especially the site where the local bistecca consumption gets closer to, and ideally meets, the product's "natural" origins; origins in terms both of geography and of the non-human living beings found at the very beginning of the production process.

The second way in which, we suggest, "nature" is present to the bistecca relates to its very appearance. Precisely, whereas other typically Florentine foodstuff, such as, for instance, the well-known *lampredotto* sandwich, entail in their appearance and presentation the necessary human labour, the latter understood as a transformative intervention onto the product, the bistecca, by being a piece of raw meat, appears divested of any intervention that goes beyond the animal's killing and segmenting. Moreover, the raw meat has a considerable visual impact on the beholder, especially when located in spatialities where it is not expected, and thus becomes "out of place". The exposition of the bistecca in the fridges-as-windows amplifies the association to the unprocessed as the meat's hanging behind the windows is considered to be a "natural process", and is advertised as such, although the meat is situated in highly calculated conditions in terms of air quality, temperature, illumination and time. Nevertheless, this "naturalness" of the bistecca as raw meat extends to its cooking and consumption: "I see nature even in the way it is consumed", the President of the Bistecca Academy stated in the interview; "It is cooked in a very simple way, on the grill, with elements that are natural (i.e., embers). One should add nothing, not even one condiment, neither salt nor olive oil".

To link to the discussion on authenticity, this silent nature embedded in the construction of the bistecca might be a significantly enhancing element to its construction as an "authentic" local product too. If "The 'authentic' is placed against the apparently artificial' (Probyn, 2003, p. 26), then it must somehow be congruent to the non-artificial, in other words to the apparently 'natural'". Consequently, the excessive visibility of the bistecca in the UNESCO Florence adds significantly to the authenticity of this specific foodstuff which, immersed in the heritagized area, acquires the quality of heritage itself. If authenticity is constructed to be consumed (Zukin, 2011), the act of ingestion amplifies this consumption. The following section provides a bistecca-focused analysis of reviews that illustrate how the exposure and specific visuality of raw meat get entangled with ideas of atmosphere and heritage so as to produce a satisfying touristic experience. We are aware that the following analysis is by no means exhaustive as, equally, is not exhaustive an exclusive focus on restaurant reviews if one seeks to investigate the role and implicit instrumentalization of nature in the process. It is nevertheless important to consider the performativity of online reviews as well as their agency in reproducing and reinforcing the touristic city, given their capacity to reach a wide public, that often base their consumption decisions on them.

The Bistecca experience. The analysis of clients' reviews in five restaurants in Florence's heritage area

In order to analyse how nature is incorporated and instrumentalized in the heritage-making machine, this study presents a web-based investigation of reviews of a selection of restaurants in Florence's UNESCO area. As known, in the contemporary landscape of the restaurant industry, the burgeoning significance of online reviews profoundly shapes consumer decisions and the standing of eateries (Hyde, 2014). At the same time, it has become evident that writing a review is not merely a simple evaluation of a culinary experience but contributes to generating discourse with profound social and spatial implications. As Zukin et al. (2017) have shown, posting a review is an exercise in presumption, where users co-produce the content they simultaneously consume. Indeed, as online review platforms provide a geographic organization of reviews, the contents generated by reviewers extend far beyond a restaurant's physical walls (Zukin et al., 2017), impacting the area (the neighbourhood, the district, and the city) in which the restaurants are located. On these platforms, reviewers conceptualize and represent the urban environments of restaurants, thereby contributing to the broader narrative concerning the culinary landscape and cultural identity of the locality. In this process, review platforms serve as a "discursive space" (Zukin et al., 2017) where local culture and identity converge, reflecting consumer preferences.

Within this framework, the study presents a comprehensive analysis of 965 reviews of five restaurants in Florence's UNESCO area, sourced from TripAdvisor and Google review platforms.[1] All five restaurants included in the research are characterized by the display of meat in their windows and inside the premises, and more generally present themselves as destinations specialized in the preparation of steak. Moreover, they are those who target explicitly an international audience and, thus, we have identified as appropriate in order to interrogate the tourism-heritage-authenticity nexus. For each restaurant, the reviews were analysed chronologically, starting from the most recent, and in various languages, Italian included. The simultaneous translation service into English provided by the platforms allowed the use of reviews in languages unknown to the authors. Although a clear distinction between residents and tourists is not possible, the analysis of the reviews reveals how the vast majority of reviews were presumably written by the latter, precisely because the culinary experience is often mentioned in the context of the broader city visit. For this reason, although the distinction between tourists and residents is complex and not straightforward, given that both groups are often involved in shared consumption habits, here the reviews are referred to as an expression of the perspective of temporary visitors.

Employing a critical discourse analysis methodology, each review underwent meticulous categorization, with distinct labels assigned to facilitate an assessment of contents. This method and analysis allow us to navigate through diverse facets of the customer experience, revealing the discursive landscapes surrounding the steak, and contributing to the atmospheres of its consumption. Words and phrases in quotation marks are taken directly from the reviews.

First, the reviews reveal how the *bistecca* plays a pivotal role in the visitors' expectations regarding the spectacularized heritage urban experience: the steak is labelled as the "star of the show", a "shining star", and the "outstanding steaks" are what make the experience "unforgettable" and "unique". The reviews show in the first place how the steak is directly associated with the concept of authenticity. Adjectives such as traditional, typical, local, Italian and Tuscan are remarkably recurring, and often accompanied by the idea that tasting the *bistecca* is a crucial component of what makes the tourist experience so unique. "Do not miss the opportunity to try unique flavours"; "it offers the recipes of the local tradition in a special way"; and "If you're looking for a good Italian experience, this is the right place" are examples of the discursive components that associate the Fiorentina with the concept of authenticity. "A steak lover paradise", "exceptional Florentine steak" and "legendary Florentine steak" are further examples of how Bistecca, authenticity and the exceptionality of the visit experience intersect one with another.

For this chapter's aims, it is useful to delve into the mechanisms that facilitate the attribution of the authentic category to the *bistecca* and the related tasting experience. It is in these mechanisms, in fact, that the appeal to naturalness comes into play. The concept of authentic is explicitly anchored to two elements closely linked to each other: the quality of the product; and the atmosphere of the place. Generally, the tasting experience of the steak is appreciated because it is perceived and recognized as of quality. In this recognition, information about the properties of the meat that is exposed and consumed is fundamental. A wide range of reviews mention the "Chianina" bovine breed as a qualifying aspect of the venue's offering: "their main attraction is the *Bistecca alla Fiorentina di Chianina*". In general, what is appreciated most is the perception of transparency and clarity of information regarding the origin and properties of the meat being consumed. In fact, it can be argued that in a widespread imaginary the Chianina breed is highly associated to the "typical" Tuscan rural landscape, such as that surrounding the city of Florence, and thus external to, but not distant from it. The (imaginary and practical) positioning of the Chianina there shapes an important, yet often implicit, link to "nature".

In this perception, the visibility and general staging of raw meat come decisively into play as a relevant factor: "The meat aging process is explained"; "They have dry age cells"; "We were enticed to enter by the beautiful pieces of meat in the window in a dry ager"; "There was a beautifully aged ribeye in a dry ager"; "I saw their meat in the window. It looked like some of the best-aged meat I have ever seen". The visibility of the steak is seen as an indicator of its quality. It is not a coincidence that restaurateurs facilitate the recognition process: in at least one of the most famous venues to taste the *bistecca*, there are labels indicating the properties, places of origin and history of the Chianina breed, and on the menu, cuts and different breeds are clearly distinguished, also in terms of price, resulting in what Kingler et al. have called "socio-material assemblage of culinary heritagization" (2023). Conversely, the absence of information is seen by many consumers as a serious gap, and as what distinguishes an "authentic" place from a "tourist trap": "Worth noting: we were told the Bistecca promotional menu had Polish rather than Florentine beef, save yourself the money and go across the river with the locals" is the suggestion of a disappointed tourist, for example.

In addition to quality, the analysis shows also how the atmosphere contributes to building the category of "the authentic" in consumers' perception. The environment, design and service all become factors that are taken into consideration so as to evaluate how a place is "the perfect place if you want to eat real Tuscan cuisine". Even in this case, the visibility of the meat plays a prominent role in conveying a "special atmosphere": "plus, it has amazing visual appeal", "awesome meats", "the appearance is amazing" are some of the recurring definitions that define how meat creates a culinary affective atmosphere "too good to be true". Some reviews focus precisely on the aesthetic dimension of the display: "the meat looked absolutely delicious to the eye"; "it arrived with a great scenic effect".

If we consider nature in terms of a social construction functional to certain (economic and political) objectives and socio-cultural discourses, and not as an objective and intrinsic quality of an object or a certain reality, the above brief analysis illustrates a highly selective process of constructing the naturalness of the meat displayed in the shop windows of Florence's UNESCO area. The staged meat is primarily functional to convey the idea of authenticity of the experience, and as such is perceived and viewed by consumers: indeed, in the long list of reviews analysed, references to the animal as a living being or biological entity were hard to find. Instead, the meat represents a component of the local cultural heritage, evaluated according to certain aspects of its naturalness (breed, origin, aesthetics) that allow associating the idea of authenticity with those of quality and atmosphere, building the fundamental traits of a unique tasting and tourist experience that satisfies the tourist's search for distinction within a heritagized urban space to be ingested (on the relation between foodification and distinction, see Bourlessas et al., 2021).

Conclusions

This chapter has sought to situate the concept of "heritage-making machine" (Gravari-Barbas, 2018) in the overtouristified UNESCO area of the Tuscan capital by centring on a particular phenomenon that has been reshaping the local retailscape in the very last years, that is, the exhibition of the *Bistecca alla fiorentina* famous steak in the windows of a considerable number of specialized restaurants (or promoted as such). Instead of approaching this practice simply in aesthetic terms, we have suggested that the produced proliferating, and arguably remarkable, visuality of raw meat is essential to the re-stabilization of the tourism-heritage-authenticity nexus, which might get destabilized in cases of overtourism where the already established tourism-related territorial fixes are threatened (Gibson, 2019). Indeed, in the same framework, scholars have highlighted that overtourism might also threaten the perceived authenticity of places (Rickly, 2019). When threatened, it has to be somehow re-established; and especially when it is loaded with heritage, the need of re-establishment might appear urgent.

From our perspective, raw meat is a successful and convincing element in the discussed heritage-making machine due to its strong, although often implicit, associations with nature; through the excessive and well-curated visibility of the bistecca, "nature" (or more precisely, external*ized* nature, following Castree, 2003) becomes what we have called the "silent" key that manages to keep the staging of authenticity together. We suggest that it is silent precisely because it has the capacity to generate implicit entanglements with "nature", the latter in a reinforcing relation with authenticity, given that both notions presuppose, and seek to stabilize, a relation with a backward space-time. Additionally, in its crudeness, the exposed meat generates an aesthetic performativity that is difficult to be achieved by any other local product, highlighting also that the tourism-authenticity-heritage imaginary nexus might also depend on exercising—and visually emphasising—power and control over non-human bodies. Through a critical discourse analysis of online reviews of restaurants that expose the bistecca in the ways that are of interest to us, we have sought to illustrate that the new staging practice that massively attracts the visually hungry tourist gaze is not just visual: instead, it is essential to the construction of atmospheres of authenticity that restabilize the so fragile tourism-heritage-authenticity nexus, constructing thus the conditions of consumption in which the desired tourist experience is eventually satisfied.

Note

1 The analysis of the reviews presented here was conducted with the assistance of Matteo Bordacconi, a post-graduate student in *Geography, Spatial Management, Heritage for International Cooperation* at the University of Florence. This work has been conducted within an internship program at the *Laboratory of Social Geography* in the Department of History, Archaeology, Geography, Fine and Performing Arts.

References

Amore, A., Falk, M., & Adie, B. A. (2020). One visitor too many: Assessing the degree of overtourism in established European urban destinations. *International Journal of Tourism Cities*, 6(1), 117–137. https://doi.org/10.1108/IJTC-09-2019-0152

Bourlessas, P., Cenere, S., & Vanolo, A. (2021). The work of foodification: An analysis of food gentrification in Turin, Italy. *Urban Geography*, *43*(9), 1328–1349. https://doi.org/10.1080/02723638.2021.1927547

Büscher, B., & Fletcher, R. (2014). Accumulation by conservation. *New Political Economy*, *20*(2), 273–298. https://doi.org/10.1080/13563467.2014.923824

Castree, N. (2003). Commodifying what nature? *Progress in Human Geography*, *27*(3), 273–297. https://doi.org/10.1191/0309132503ph428oa

Celata, F., & Romano, A. (2020). Overtourism and online short-term rental platforms in Italian cities. *Journal of Sustainable Tourism*, *30*(5), 1020–1039. https://doi.org/10.1080/09669582.2020.1788568

Falconer, E. (Ed.). (2021). *Space, Taste and Affect. Atmospheres That Shape the Way We Eat*. Routledge.
Gibson, C. (2019). Critical tourism studies: New directions for volatile times. *Tourism Geographies*, *23*(4), 659–677. https://doi.org/10.1080/14616688.2019.1647453
Gravari-Barbas, M. (2018). Tourism as a heritage producing machine. *Tourism Management Perspectives*, *25*, 173–176. https://doi.org/10.1016/j.tmp.2018.01.004
Hyde, Z. (2014). Omnivorous gentrification: Restaurant reviews and neighborhood change in the downtown eastside of Vancouver. *City & Community*, *13*(4), 341–359. https://doi.org/10.1111/cico.12088
Klingler, M., Schermer, M., Hemetsberger, A., Stotten, R., & Maaß, C. (2023). Uncovering the sociomaterial assemblage of a culinary heritagization: The Wildschönauer Krautinger schnapps. *Journal of Rural Studies*, *103*, Article 103125. https://doi.org/10.1016/j.jrurstud.2023.103125
Liberatore, G., Biagioni, P., Ciappei, C., & Francini, C. (2022). Dealing with uncertainty, from overtourism to overcapacity: A decision support model for art cities: The case of UNESCO WHCC of Florence. *Current Issues in Tourism*, *26*(7), 1067–1081. https://doi.org/10.1080/13683500.2022.2046712
Loda, M., Bonati, S., & Puttilli, M. (2020). History to eat. The foodification of the historic centre of Florence. *Cities*, *103*, Article 102746. https://doi.org/10.1016/j.cities.2020.102746
Mariyani-Squire, E. (1999). Social Constructivism: A flawed debate over conceptual foundations*. *Capitalism Nature Socialism*, *10*(4), 97–125. https://doi.org/10.1080/10455759909358888
Philo, C., & MacLachlan, I. (2018). The strange case of the missing slaughterhouse geographies. In S. Wilcox & S. Rutherford (Eds.). *Historical animal geographies* (pp. 86–106). Routledge.
Probyn, E. (2003). *Carnal appetites: Foodsexidentities*. Routledge.
Puttilli, M., Bonati, S., & Portinaro, L. (2020). Urban meatification. Esplorazioni visuali tra tutela e consumo del patrimonio culturale nel centro storico di Firenze. *Geotema*, *62*, 61–71.
Puttilli, M., & Loda, M. (2022). History (still) to eat: l'impatto della pandemia sul settore della ristorazione nel centro storico di Firenze. *Rivista geografica italiana: CXXIX*, *4*(2022), 13–29.
Rickly, J. M. (2019), Overtourism and authenticity. In R. Dodds, & R. W. Butler (Eds.), *Overtourism. Issues, realities and solutions* (pp. 46–61). De Gruyter Oldenbourg.
Urry, J. (2002). *The tourist gaze*. Sage.
Zukin, S. (2011). Consuming authenticity: From outposts of difference to means of exclusion. In S. Binkley, & J. Littler (Eds.), *Cultural Studies and Anti-Consumerism* (pp. 206–230). Routledge.
Zukin, S., Lindeman, S., & Hurson, L. (2017). The omnivore's neighborhood? Online restaurant reviews, race, and gentrification. *Journal of Consumer Culture*, *17*(3), 459–479. https://doi.org/10.1177/1469540515611203

Part II

Narratives of nature in the contemporary environmental crisis

6 The vanishing shapes of Dolomitic ice

Temporalities and aesthetics of Marmolada glacier on the threshold of its extinction

Mauro Varotto and Matteo Vianello

Introduction

This contribution aims to examine the critical situation of temperate alpine glaciers from a perspective that combines cultural and glaciological dimensions, a specific declination within the broader historical debate on the relationship between historical and biological/ecological times (Tiezzi, 2001). The reflection addresses a re-examination of the relationship between humans and nature in the era of climate change, taking inspiration from one of the most celebrated mountain groupings in the Dolomites and the eastern Alps, Marmolada, a UNESCO World Heritage Site since 2009. Marmolada is a "symbolic mountain", well known for several reasons, in addition to being the most important glacier in the Dolomites:

- From an aesthetic standpoint, together with Tre Cime di Lavaredo, it is the most iconic and frequently depicted peak in the Dolomites.
- From a historical standpoint, for at least a century, its glacier has been among the most popular and "infrastructured" in the Alps, starting from the *Eisstadt*, the "City of Ice", 12 kilometres of tunnels interspersed with caves used as dormitories, kitchens, radio rooms, and canteens dug by Austrian soldiers into the belly of Marmolada glacier during the First World War.
- From the standpoint of climate change, it became tragically famous around the world for the ice collapse of 3 July 2022, in which 11 people lost their lives.

Marmolada glacier will be among the first "famous" glaciers in the Alps to disappear, most likely well before 2050 (Varotto & Ferrarese, 2021). For all these reasons, it is a particularly significant mountain for observing the evolution of relationships between humans and glaciers, between natural dynamics and cultural attitudes, and for analysing the "temporalities" that affect the glacial landscape in its material and immaterial dimensions.

From observations of the varying speeds in the processes that have affected the glacier in the recent past, we have organised this reflection into three parts corresponding to "temporal perceptions" in the reading and interpretation of

DOI: 10.4324/9781003465164-9

the glacier—a first "slow" almost immobile period, reflected both in glaciological dynamics in the strict sense and in the aesthetics of the glacier; a second more paroxysmal and "emergency" period prefigured by the acceleration of the processes of exploitation of the glacier in the years of the tourism economic boom and by the recent melting process of the glacier, culminating in the tragedy of 2022; and finally, the prefiguration of a "third period", which in the last five years, even before the tragedy of 2022, presents itself as an attempt to join previously distinct temporalities. We could also define the latter as the "time of the Anthropocene", in which the human dimension is called upon to take charge of the *longue durèè* of post-glacial dynamics (Braudel & Wallerstein, 2009)—a temporal reconciliation that requires abandoning the modernist approach and focusing on the idea of the glacier as both a scenically inert entity and a fluid reality "to be saved".

An essential reference for the reflections of the chapter is the essay *Temporality of the Landscape* by anthropologist Tim Ingold (1993), in which the author suggests a cartographic, archaeological, and anthropological approach that reads the territory as a palimpsest, thus activating conceptual bridges between different times and spaces. The landscape is thus configured as a "synthesis of different temporalities" (Schama, 1997; Corboz, 1983), and as such, a hybrid entity in which continuous interaction between human and ecological dimensions takes place, crossing and undermining the nature–culture binary, a dualism that is perhaps at the origin of contemporary ecological crises. If the landscape lends itself to being that bridge concept in which the relationality between observer and observed reality, between subject and object finds space and meaning (Farinelli, 2003; Illetterati, 2023), the glacial landscape can be considered an archive that is both human and non-human, offering a stratigraphy capable of recording within it not only different glacial dynamics but also different aesthetic perceptions and uses that have occurred over time. In conclusion, the very notion of time—understood as a homogeneous and universal unit of measure—is abandoned in favour of the more complex one of "temporality" (Ingold, 1993), given by the relationality between the becoming of things and their conceptualisation.

First period: Marmolada "still" and sublime

The "first period" of Marmolada has its roots in the Ladin legend of its genesis (Mari & Kindl, 1989), a sort of divine curse generated by an act of *hubris*: failure to respect the rest of times, both religious and natural, by an old peasant woman who broke the holiday obligation to collect more hay; divine punishment, in the form of endless snowfall, covered the green slopes of Marmolada with ice, transforming it into a perennial warning of broken balance. The legend places the origin of the glacier within a mythical and timeless dimension, a local declination of the myths of the Golden Age, which, from the Great Flood back to expulsion from the Biblical Eden, attribute the crisis of an Edenic situation to a human infraction punished by divinity. The glacier, in its immobility, becomes something supernatural that imposes itself on the body of an old woman guilty of greed and becomes a warning destined to last forever. The time of the glacier is therefore, from its origins in a time "other" than human, a "divine" space–time that represents an insurmountable physical and moral boundary for humans.

The glacier was also an "other" dimension for shepherds who, since ancient times, frequented the rich pastures of Passo Fedaia, the plateau overlooking Marmolada from the north. Until at least the eighteenth century, there was no relationship, attendance, or toponymic trace relating to the glacier, with the exception of the names Sasso delle Undici and Sasso delle Dodici ascribed to the rocks emerging as natural sundials from the sea of ice. This immobile eternity reverberated on other levels and for other audiences with the affirmation of the Grand Tour and the mountaineering discovery of the Dolomites in the nineteenth century. Since then, the image of the glacier has been reproduced and multiplied infinitely in drawings, pictorial representations, photographs, and the "postcard landscapes" of the nascent tourist epoch (Castree, 2003). It was a distinctive presence, a point of reference immobile in its whiteness, in the same way that, from the Venetian plain, its whitewashed peak appears as a timeless background capable of providing a stable framework and a fixed point of reference to the dynamism of the urban territories of the lowlands (Gagné et al., 2014).

This dichotomy and aesthetic divergence are at the origin of a political and socioeconomic contrast between two worlds increasingly conceived as separate—the nature of the glacier extends to the surrounding mountains, thus imprisoned in mountain-eternity-tradition-wilderness, as opposed to plain-progress-mobility-modernity (Bätzing, 2005; Camanni, 2002; Varotto, 2020). The immobile and "other" dimension of the glacier is, therefore, a figure of a broader process of marginalisation and expulsion from the human sphere of a mountain increasingly conceived as Nature, a sacred and inviolable place, therefore naturalistically protected and culturally patrimonialised (De Rossi, 2014), and by virtue of this, it has become an attraction, a destination, a tourist commodity—a colonial vision that becomes a factory of imagery for the benefit of consumers in their leisure. The cursed ice of legend is thus transfigured and becomes the white mantle of the "Queen of the Dolomites", as Marmolada is locally called, which will be conquered by mountaineering in 1862 by Paul Grohmann.

The mountaineering conquest preceded glaciological observations by a few decades, which continued from the end of the nineteenth century throughout the twentieth century, making Marmolada both a scenic backdrop and one of the most studied glaciers in the Alps (Carton & Varotto, 2011). At least four generations of researchers took turns taking measurements on the glacier, beginning with Giovanni Marinelli, one of the fathers of Italian geography (Mattana, 1995). This first phase of glaciological research contributes to translating the "eternal" time of the glacier into "very slow movement"—the imperceptible movements of glacial masses, their variation from year to year, the complex dynamics between the atmosphere and the cryosphere that generate minute movement—and translates the immobility of the glacier into a dormant river with surface oscillations and frontal measurements for at least half a century, far from the accelerated melting of the last 30 years. These studies reveal the slow movement of the glacier, as will also be confirmed by the testimonies of the Austrian soldiers who built the tunnels of the *Eissstadt*, the "City of Ice" in 1916 (De Bernardin & Wachtler, 2009), which, however, never question the overall stability of the glacial system, crossed by the "balance line" that separates the supply basin and the ablation basin.

Second period: The great acceleration and the emergency phase

Meanwhile, in this apparently inert or extremely slow-moving phase until the first half of the twentieth century, other more rapid and aggressive temporalities appear, which imprint a clear change in speed after the Second World War compared to the slow rhythms of the glacier: its "otherness" (Aime & Papotti, 2012) is progressively colonised and conquered, first with the installation of the Graffer brothers' ski lift in 1947, one of the first in the Alps, then with the SADE hydroelectric project and construction of the Fedaia Dam, inaugurated in 1954, and finally with the construction of the first section of cable car between Malga Ciapela and Serauta in 1967. These are the stages of an appropriation process destined to trigger a 30-year (1972–2002) border dispute between Veneto and Trentino for economic control of the glacier, which had become "white gold" for the ski industry.

Climate change, in turn, has rapidly accelerated the melting dynamics of the glacier since the 1980s. The "balance line" has disappeared, positioning itself almost permanently beyond the summit levels of the glacier, which lost over 80% of its surface and volume in the last century (Varotto & Ferrarese, 2021). Summer skiing entered a crisis and ceased to be practiced in 2005, while winter skiing persisted, supported by artificial snow systems and snow farming. On 3 July 2022, this acceleration was affected by an unpredictable paroxysmal episode: a portion of Marmolada glacier suddenly detached from the central body at over 3000 metres above sea level, below Punta Rocca, triggering an avalanche of 65,000 cubic metres of ice and debris that collapsed at a speed of 150 km/h, overwhelming 11 climbers on the normal route to the summit. In live video recordings of the event, the dust raised by the avalanche completely fills the space of the frame. After a few hours, the first photographs taken by an air ambulance depict the scar left by the detachment of ice—a chasm at 3250 metres above sea level with a 25-metre overhang exposes the entire internal structure of the glacier, revealing its stratifications. From the white of the surface snow to the bright blue of the ice to the dark grey of the basal layer in contact with rock, its composite being is laid bare, the sum of many temporalities. It is a gash that dissects and ruptures forever our age-old trust in the slowness of the glacier's advance.

The news appeared in national and international media, and Marmolada once again became a global icon. The "tragic" event took on tones of a "catastrophe" or an etymologically perceptive "reversal" of the immobile glacier. It inaugurated a change of posture concerning the interpretation of the timing (and predictability) of avalanche events but also of the glacier as a whole and of climate change on a larger scale. The glacier had suddenly sped up, as we did. The "first period", that of a clear separation between the human and the ecological spheres, definitively ended. The avalanche, unprecedented in proportions and dynamics, was a clear demonstration that the point of equilibrium in the glacier no longer existed, and that glaciological and human practices had spilled over into each other. From the standpoint of the relationship between humans and the glacier, a temporal passage opened—the period of emergency was inaugurated due to the collapse of not only spaces but times, which, in spite of themselves, found themselves compressed to such an extent as to be superimposable on the acceleration of the temporalities of human activity ("Marmolada of Madness|Not|NERO", 2022).

In the scenario made explicit by the Marmolada tragedy, the space–times linked to the rapid retreat of the glacier exacerbated the contrast between the historical persistence of the ice landscape and its gradual dissolution. In this key, the images of the avalanche disavow the temporal structure assigned to the glacier—that of a cosmic and immutable time in which the landscape does not move. This is a new scenario for the representation of the mountain and its temporality. Before the Industrial Revolution and the tourist epoch, Marmolada glacier was set in a cosmic time suspended compared to that of lower lands. The mountains figured, as in the Western landscapes described by Deleuze (Deleuze et al., 1986), as a naturalistic backdrop that founded and predestined urban civilisations. From the lagoon of Venice—in the past, the political power that dominated the Dolomites—the mountains had always been a still backdrop, enveloped in an archaeological time incommensurable with the dynamism of the city. With the conquest of the mountain in the contemporary age, and almost leveraging this immobility, urban man has taken possession of the glacier, imposing upon its apparent stability his roads, dams, and ski lifts, which allowed access to the mountain and consumption of its landscape, drawing it into his urban time. The sublime and immutable time of the Dolomites thus became a function of the tourist industry, which immediately after the tragedy demanded the safety of the glacier. What mattered, first and foremost, was to overcome the temporal short circuit, guarantee stability for future tourist seasons, and lift the ban on access to Marmolada as soon as possible.

There is no doubt that the major change induced by the avalanche is not so much the morphological one but that linked to the mechanisms of mediatisation of the glacial environment: the radical acceleration of climate catastrophes is reverberated in the rapidity with which the images spread, a dynamic of whirlwind, and very fast "spectacularisation" (Debord & Stanziale, 2008). The avalanche inaugurated the time of the "dying glacier", an image of powerful humanisation also shared by international literature, which defines the planet's temperate glaciers as "endangered species". Despite sharing a parallel history with alpine history, the glacial landscape deserves a specific discussion; these are ecologies whose transformation is now foreseen as irreversible such that they are conceived as close to their definitive disappearance. While the humanisation of the glacier materialises in the aesthetics of art and the performance of funeral rites (such as the funeral song promoted by the Glacier Caravan: Legambiente, 2020), the glacier today is suspended in a temporal diaphragm between the no-longer-as-before and the not-yet-disappeared (Dematteis & Nardelli, 2023). As in the Doomsday Clock—an imaginary clock that estimates annually the minutes remaining until the end of time—we perceive the time of Marmolada as an asymptote that points to the final date without ever reaching it—an immobile transition between now and the imminent death of the glacier, which, however, in its relatively slow agony, has not yet made the news.

The emergency thus becomes the aesthetic and media lens capable of representing—and at the same time mystifying—the relationship between human and glacial melting: the glacier becomes human, and we lose sight of the relationship and the interweaving of different temporalities, the relationship with otherness. The rhetoric of the "glacier to be saved" justifies the operation of covering the ski slope with geotextile sheets, which began in 2005. To date, approximately 40,000 square metres of sheets are spread annually on the glacier (a coverage that affects just 3% of its surface, corresponding to the route of the sky run). The sheets do not actually save the glacier, but only its image for public opinion and the tourist season, in the event of poor snowfall or even in the event of the disappearance of the glacier itself. A second short circuit is therefore generated: the human activity indirectly responsible for the eventual disappearance of the glacier is perpetuating itself in the name of saving the glacier and continuing its work of destruction.

The installation of the geotextile sheets appears, at first sight, as the logical realisation of the glacier's emergency period. Faced with its rapid dissolution, these sheets will make it possible to slow down the fusion process at least immediately. However, the use of this technique neglects a long-range vision in multiple ways: geotextile sheets are very expensive (0.6–0.8€ per cubic metre of ice), they introduce pollutants into the glacial ecology (from microplastics to carbon dioxide emissions generated by their installation), and on an ecological level, they become a real barrier in the glacial mass (Comitato Glaciologico Italiano, 2022). The result is once again the apparent crystallisation of glacial times in favour of human times.

Third period: Anthropocene, prediction and responsibility

The "second period", the time of emergency on the emotional wave of the tragedy, generated two reactions of opposite nature. On the one hand, the "business as usual" approach, which considered the tragedy an isolated episode, tends to minimise its revelatory potential and seeks to guarantee the immutability of the glacier and human action on it. In other words, humans and nature remain separate realities, incapable of deep dialogue and mutual recognition. On the other hand, a different awareness is finding its way—the sensation of a veil being torn away by the collapse between historical times and glaciological times, and the need to rethink the relationship with the glacier and the mountain (Varotto, 2020). This new approach to the glacier starts from the rethinking of humanity, which is reflected in, and the spokesperson for a different approach to the mountain in its entirety, a prelude to a "third period", that of the Anthropocene.

The collapse of the ice wall breaks distances and opens a new proximity: the glacier has not yet completely descended into the valley, but the image that can be perceived from the plain is far from that of an unchanged and immutable background. The immediate time of the tragedy raises the question of what the ice's true time is—its old age, its life expectancy, and the acceleration of its transformations. The scientific debate following the tragedy brought scientific studies and cultural stances to light that have warned for years of the precarious state of the ice landscape. The third period materialised, even before the tragedy, in the repeated alarms of glaciologists and their exhortations to take a step back (Bonesio, 2011) in the participatory glaciological campaign and citizen science initiatives launched by the Italian Glaciological Committee and the Geography Museum of the University of Padua since 2019, and finally, in the *Marmolada 2020 manifesto: for one sustainable development* launched by Guido Trevisan, manager of the local Pian dei Fiacconi refuge. This manifesto aims to give Marmolada a new symbolic meaning: no longer the avant-garde of tourist modernity but an alternative and anticipatory model of more sustainable development capable of taking charge of different temporalities in the Anthropocene era. It is an invitation to find a point of mediation between human times and the times of the glacier, slowing down our pace, abandoning the ski season in favour of more sustainable practices, and planning cultural and awareness-raising routes towards the issues of climate change.

The catastrophic event of July 2022 highlights the temporal short circuit between the gears of modernity. Through the "heuristic of fear", Jonas indicates tragedy as a fundamental element for recognising human responsibility towards future generations and towards the glacier.

> Only the expected upheaval of man helps us to formulate the relevant concept of humanity to be safeguarded; we need the threat of human identity—and absolutely specific forms of threat—to make us anxiously ascertain the real identity of man.
>
> (Jonas, 2014, p. 35)

The third period imposes not only a dialogue between different temporalities but also a broadening of the relational horizon between what happens on Marmolada and the broader economic and cultural context. The glacier is therefore no longer an immobile background, but rather a thermometer and a test capable of inspiring more responsible human actions, an ethics based on multiple temporalities that overcome the emergency, itself the daughter of simultaneity and immediacy. In this sense, Marmolada, from a forerunner of the extractive logic of modernity, becomes a glacier-laboratory precisely because of its extinction, inspiring an ethic of prediction and responsibility challenging humanity to rethink itself and "become landscape" (Illetterati, 2023), in other words, to orient political-economic actions open to future generations and non-human entities, definitively dismantling the image of the mountain as an inert space, a stage backdrop, or an eternal curse.

References

Aime, M., & Papotti, D. (2012). *L'altro e l'altrove: antropologia, geografia e turismo.* Einaudi.

Bätzing, W. (2005). *Le Alpi. Una regione unica al centro dell'Europa.* Bollati Boringhieri.

Bonesio, L. (2011). In morte di un ghiacciaio. In A. Carton, & M. Varotto (Eds.), *Marmolada.* Cierre edizioni.

Braudel, F., & Wallerstein, I. (2009). History and the social sciences: The Longue Durée. *Commemorating the longue durèe*, *32*(2), 171–203.

Camanni, E. (2002). *La nuova vita delle Alpi.* Bollati Boringhieri.

Carton, A., & Varotto, M. (2011). *Marmolada.* Cierre edizioni.

Castree, N. (2003). Commodifying what nature?. *Progress in Human Geography*, *27*(3), 273–297. https://doi.org/10.1191/0309132503ph428oa

Comitato Glaciologico Italiano. (2022). Coprire i ghiacciai con teli non significa salvarli. https://www.planetmountain.com/it/notizie/ambiente/coprire-i-ghiacciai-con-teli-non-significa-salvarli.html (Accessed: 3 March 2023).

Corboz, A. (1983). The land as palimpsest. *Diogenes*, *31*(121), 12–34. https://doi.org/10.1177/039219218303112102

De Bernardin, A., & Wachtler M. (2009). *La città di ghiaccio. La Grande Guerra nelle viscere della montagna.* Athesia.

De Rossi, A. (2014). *La costruzione delle Alpi: immagini e scenari del pittoresco alpino (1773–1914).* Donzelli.

Debord, G., & Stanziale, P. (2008). *La società dello spettacolo.* 3rd ed. Massari.

Deleuze, G., Tomlinson, H., & Habberjam, B. (1986). *Cinema 1: The movement-image.* University of Minnesota Press.

Dematteis, M., & Nardelli, M. (2023). *Inverno liquido La crisi climatica, le terre alte e la fine della stagione dello sci di massa.* DeriveApprodi.

Farinelli, F. (2003). *Geografia. Una introduzione ai modelli del mondo.* Einaudi.

Gagné, K., Rasmussen, M. B., & Orlove, B. (2014). Glaciers and society: Attributions, perceptions, and valuations. *WIREs Climate Change*, *5*(6), 793–808. https://doi.org/10.1002/wcc.315

Illetterati, L. (2023). Diventare paesaggio. Prospettive ontologiche, pratiche e politiche. In *Studi filosofici.* XLVI. Bibliopolis.

Ingold, T. (1993). The temporality of the landscape. *World Archaeology*, *25*(2), 152–174.

Jonas, H. (2014). *Il principio responsabilità: un'etica per la civiltà tecnologica.* 3. rist. Edited by P. P. Portinaro. Einaudi.

Legambiente. (2020). Carovana dei ghiacciai, tappa Marmolada: i risultati del monitoraggio sul ghiacciaio • Legambiente. *Legambiente*, 29 August. https://www.legambiente.it/notizie-dal-territorio/carovana-dei-ghiacciai-tappa-marmolada-i-risultati-del-monitoraggio-sul-ghiacciaio/ (Accessed: 8 February 2023).

Mari, A., & Kindl, U. (1989). *La montagna e le sue leggende.* Mondadori.

Marmolada of Madness|Not|NERO. (2022) *Not.* Available at: https://not.neroeditions.com/marmolada-of-madness/ (Accessed: 13 January 2024).

Mattana, U. (1995). Il ghiacciaio della Marmolada: aspetti geomorfologici della fronte centrale. *Rivista Geografica Italiana, 102*, 113–127.
Schama, S. (1997). *Paesaggio e memoria*. Mondadori.
Tiezzi, E. (2001). *Tempi storici tempi biologici. Vent'anni dopo*. Donzelli.
Varotto, M. (2020). *Montagne di mezzo. Una nuova geografia*. Einaudi.
Varotto, M., & Ferrarese, F. (2021). Esperienze di contrasto e mitigazione dei cambiamenti climatici: il ghiacciaio della Marmolada. *Ambiente & Sviluppo*, 12 (pp. 1–35). IPSOA – Wolters Kluwer.

7 The commodification of mountain natures in the eco-climate crisis

The Milano Cortina 2026 Olympics and the Dolomites

Andrea Zinzani

Introduction

In the framework of the critical analysis of Italian socio-natures in the contemporary eco-climate crisis, this chapter seeks to shed light on commodification processes of mountain natures in the Eastern Italian Alps. Indeed, it adopts a political ecology perspective to focus on the Dolomites, analyzing the Milano Cortina 2026 Olympics' politics, visions, and narratives of mountain futures as expressed through infrastructural development projects driven by governance actors at multiple scales. Over the past three decades, critical geography scholarship has contributed significantly to understanding how the consolidation of neoliberal capitalism has shaped and reconfigured nature on different levels. Indeed, by taking on Neil Smith's (1984) argument about the production of nature, critical geography has advanced the theoretical development of socio-natures; at the same time, it has furthered our understanding of nature commodification processes through concepts such as "accumulation by conservation" (Büscher & Fletcher, 2020; Castree, 2003; Castree, 2014; Loftus, 2017).

In the framework of today's global eco-climate crisis, it is crucial to delve more deeply into controversial politics of sustainable and green nature capitalization and support alternative socio-environmental futures – such as grassroots collective community resource governance (Ernstson & Swyngedouw, 2019) – aimed at moving beyond the endless exploitation of nature and toward climate justice instead. Over the last few years, political ecology scholarship has questioned capitalist accumulation processes and their socio-environmental impacts as well as capital's domination of nature by adopting neo-Marxist, feminist and decolonial perspectives that have enriched contemporary critical geography debates significantly (Bryant, 2017; Perreault et al., 2015; Zinzani, 2020). Furthermore, in reflecting on today's global eco-climate crisis, political ecology thought has stressed the imperative of using action-research and political engagement to support more just socio-environmental futures and a radical transformative change in society-nature interplays (Benjaminsen & Svarstad, 2021; Kothari et al., 2019).

While political ecology has focused on diverse contexts to explore processes of nature commodification and exploitation, however, little research so far has

DOI: 10.4324/9781003465164-10

analyzed mountains, especially politics and visions of mountain areas in relation to the eco-climate crisis (Stoddart, 2013; Voicu & Vasile, 2022; Zinzani, 2023). Over the last few decades, mountain geography has concentrated on the evolution of mountain-urban areas interdependencies, highlighting socio-economic challenges as well as the importance of promoting a global mountain agenda for conservation and development (Debarbieux & Price, 2008; Fonstad, 2017). In this framework, Italian mountain geography scholarship has contributed to the understanding of mountain depopulation, marginality, and politics of sustainable development, as well as governance mechanisms and the role mountain communities play in Alpine and Apennine regions (Ferrario & Marzo, 2020; Varotto, 2020). However, few studies have engaged with mountains through the conceptual lens of the social production of nature (Bonati et al., 2021, among others) and its contested dimension; likewise, little attention has been granted to analyzing the politics and visions of mountain commodification and exploitation. These dynamics could be read as controversial processes of nature transformation that are deeply linked to the eco-climate crisis, calling for a profound reflection on mountain futures. Therefore, by applying this critical perspective to the Dolomites, this chapter shows how, despite the severe effects of the eco-climate crisis, mountain nature in the Ampezzo valley is still characterized by the reproduction of a dominant logic of commodification and exploitation through a naturalized vision of "accumulation by sustainability". However, this logic is increasingly challenged, today, by a growing counternarrative of mountain futures that aims to move beyond a no-limits commodification of mountains and instead envision a radical reconfiguration of nature-society relations.

For this research, the political ecology theoretical lens was methodologically combined with ethnographic research in the Ampezzo valley and town of Cortina, conducted through fieldwork in 2022. This introduction is followed by the next section which focuses in more depth on research methods and highlights the importance of ethnography for understanding the visions and narratives of mountain nature futures. The second section details the Ampezzo valley case-study by focusing on the Milano Cortina 2026 organizational process and its infrastructural development initiatives, combined with the *Carosello Dolomitico* project, and in particular visions and narratives of governance actors. The last section discusses the evidence in relation to the overall theoretical framework of the book.

Methodology: Ethnographic research in the Alps

This analysis of processes of mountain nature commodification in the Dolomites, and especially their visions and narratives, was conducted using ethnographic research. Through its diverse methods, ethnography enables an in-depth understanding of everyday geographies and especially a critical reflection on how socio-environmental relations are produced by the knowledge, experiences, and practices of individuals, collective subjects, and institutions

(Martini et al., 2022). Furthermore, ethnographic methods enable the researcher to engage as a subjective individual, part of the research process itself. Whereas ethnography has been employed by contemporary geographical research to investigate diverse contexts, this approach is still quite innovative in mountain research, especially in the Global North and the Alps.

This ethnographic research in the Ampezzo valley and Cortina d'Ampezzo was conducted through three fieldworks between February and November 2022. In terms of methods, semi-structured interviews and formal conversations were carried out with representatives of institutions at the district and local scales, from Belluno district to municipalities, members of natural parks committees, collective property unions, mountain, social and environmentalist associations, and experts on mountain issues and development as well as businesspeople in the mountain lift sector. In total, 25 interviews and conversations were conducted with institutional and social actors, with the meetings arranged in different settings such as institutional offices, Olympics development sites, bars, and mountain huts. Depending on the actors, these conversations and interviews focused on today's Dolomites socio-environmental challenges and the effects of the eco-climate crisis, Ampezzo valley environmental governance, and especially the Milano-Cortina 2026 Olympics' infrastructural development initiatives. Specific attention was paid to the idea of mountain futures in relation to the Olympics as a process of social production of nature in the eco-climate crisis. Furthermore, some of the actors involved in the research proved willing to allow field-surveys and visits to the Olympics infrastructural development sites as well. Therefore, ethnographic methodology plays a key role in enabling an understanding of actors' ideas, visions, positions, and interests with regard to Dolomites mountain natures, as well as the socio-political relations, power dynamics, and divergent and contested visions that will be discussed in the following section.

Milano Cortina 2026 and the Dolomites: the commodification of mountain nature

The Dolomites are part of the southeastern section of the Alps and are located in northeastern Italy; administratively, they belong to the provinces of Trentino Alto Adige and Veneto. The Dolomites environment is characterized by many peaks reaching a height of 3,343 m (the Marmolada), tiny glaciers, alpine meadows and forests, agricultural and pastural areas, valleys, and villages. The social production of Dolomites natures has been characterized over time by subsistence farming, forestry and pastoralism, often characterized by collective property regimes. Since the 1960s, it has also entailed increasing hydropower infrastructural development, the industrialization of valleys and a significant rise in tourism (Casanova, 2020). Tourism in particular has deeply shaped mountain development and the political-economic interests of several Dolomites valleys since the 1970s through ski resort infrastructural development and the construction of tourism facilities. The Ampezzo valley and

Cortina represent a key example of this development process, although in reality this area had proven attractive for elite alpine tourism tours as early as the late 1800s (Dematteis & Nardelli, 2022). Indeed, a wide-ranging public has often considered Cortina and its valley surrounded by high peaks and blanketed in alpine meadows and forests an astonishing example of mountain nature, famous for its beauty, that has succeeded in combining nature conservation and development. As the director of the Centro Studi sulla Montagna of Belluno observed, Italian and international writers and intellectuals such as Kipling, Hemingway, Moravia, and Carducci have eulogized the astonishing beauty of Cortina's meadows, forests, and peaks. With regard to nature in the Cortina valley, a member of Regole d'Ampezzo – the historical collective property-owning union that has controlled the majority of the valley's forests, pastures and high meadows for centuries – stated that grassroots, community conservation vision has been able to balance nature conservation and development for decades and thus prevent the massive urbanization of mountain slopes.

Since the 1980s, however, Cortina has risen to become the Dolomites' most exclusive mountain ski resort destination as a result of massive public and private financial support and its success in attracting Italian and European urban elites (Ferrario & Marzo, 2020). Indeed, Italian mainstream popular movies since the 1980s have featured Cortina as the setting for high-end, consumeristic holidays by Italian political and economic elites, showcasing Ampezzo's mountain facilities and ski slopes as the "place to be". The uniqueness of the Dolomites' mountain nature was formalized in 2009 when this range was declared a UNESCO World Heritage Site, followed by the establishment of the Fondazione Dolomiti UNESCO.[1] The Fondazione, a non-profit organization, was established by provincial institutions together with the UNESCO secretariat to govern the conservation and valorization of the Dolomites natural heritage, a patrimony increasingly threatened by the effects of climate change.

Over the past two decades and especially in recent years, the impact of climate change on alpine mountain nature has indeed intensified considerably. A 2022 IPCC report stated that mountains are one of the environments most vulnerable to the effects of the global eco-climate crisis in terms of global warming, environmental degradation, and biodiversity loss (IPCC, 2022). Over the last decade, the Dolomites have been facing a significant increase in average annual temperatures leading to a decrease in snowfall and winter snow cover, especially below 1500 m, and a summertime increase in heavy rainfall, storms, heat waves, and drought, as stated by a climate expert with the Veneto forecast service. On one hand, these climate dynamics are contributing to glacier melt, snowfield reduction and water scarcity issues, particularly in the summer; on the other hand, they are driving forest degradation and biodiversity loss, especially through the recent spread of the *bostrichidae* beetle (see: *Nuovo Catasto dei Ghiacciai Italiani*).[2] Therefore, the impact of the eco-climate crisis in the Dolomites and its effects on mountain communities' everyday life have progressively given rise to a reflection on the fragile Dolomites' environment. Public debate among institutions, mountain associations, and communities has begun to revolve around the interplay of nature conservation and mountain development, economic valorization, and tourism. How should mountain communities imagine and envision mountain futures in the context of the eco-climate crisis? What pathways of continuity or political-economic change in mountain development might channel their visions?

In June 2019, Cortina accepted assignment of the 2026 Winter Olympics. Its application dossier, formalized by Italian national sport authorities with the support of the Lombardia and Veneto provincial governments, stated: "Milano Cortina 2026 will become the most sustainable Olympic games ever and they should represent a source of inspiration to change the lives of future generations". Moreover, "the Olympics will place Italian alpine region in a key position with regard to global sustainable development strategies by paying specific attention to environmental conservation and awareness; indeed, the games will reuse, or readapt, existing infrastructures".[3] This statement, shared by Olympic and political institutions, showed an apparent commitment to limiting infrastructural development and natural resource exploitation. It is important to highlight that the Olympics were assigned to Cortina at approximately the same time the World Ski Championship (held in February 2021) was being organized. As stated by a former employee of the regional department of forests and agriculture, today a member of the local political organization Cortina Bene Comune, the phase characterized by the World Ski Championship and beginning of the Milano Cortina 2026 organizational process represented a key instance of momentum: it was a chance to revamp Cortina's image through mountain politics involving the organization of international sport events supported by institutional interests and private capital flows at different scales. Indeed, the head of the mountain lift company Funivie Marmolada added that the World Ski Championship and Olympics constituted a key opportunity to make new investments in summer and winter tourism, as well as to strengthen sustainable development strategies. In reflecting on this process, the President of the committee Comitato Civico Cortina argued that the World Ski Championship organizational process in particular represented the beginning of a new wave of mountain nature infrastructural valorization, characterized by powerful actors launching projects on multiple scales and a novel – and controversial – narrative of sustainability, inclusion, and legacy. However, this valorization process also entailed environmental impacts on mountain nature, as noted in informal conversations with Cortina citizens. For instance, locals highlight that preparations for the 2021 World Ski Championship involved extending ski slopes in which portions of forest were removed, the mountainside was reshaped via excavations and a reservoir to store water for artificial snow was created. Furthermore, Olympics projects such as developing the bobsled slope and Olympic village have already involved and will continue to involve forest cutting, biodiversity reduction, river water diversion, and valley floor urban development. To understand this wave of nature infrastructural valorization and especially the politics, visions, and narratives of mountain futures, it is important to delve into the driving forces behind the Olympics organizational process and especially its main infrastructural development. With the support of the International Olympic Committee, the national government and Veneto and Lombardia provincial governments, the Fondazione Milano Cortina 2026 was established to oversee the organization of the Olympics while Società Infrastrutture Milano Cortina 2026 was founded to develop infrastructure (Casanova, 2022). These newly established entities include national experts and businesspeople as well as the representatives of provincial institutions. Through the narrative of legacy, sustainability, and inclusiveness, they have proposed three key projects (among other secondary ones) to be developed in Cortina (Figure 7.1).

The first initiative involves building a ski lift connection between the two sides of Ampezzo valley via two new lifts to link downtown Cortina with the Tofane (Socrepes) and Faloria ski resorts (Casanova, 2022; Dematteis & Nardelli, 2022). The Cortina-Socrepes connection would be built on an unstable slope that had been deliberately protected from urban development, however, as stressed by a former member of the regional forests department. The second project is building an Olympic village in Flames, on the valley floor north of Cortina – an ecologically fragile area that, as highlighted by a representative of the Regole d'Ampezzo, is characterized by hydrogeological risk.

Figure 7.1 Olympics poster in Cortina highlighting the prevailing discourse.

Source: author, 2022.

Given this risk together with the issues of nature urbanization and biodiversity loss, the project has been debated and changed several times over the last two years; it is likely that the planned permanent structure will be replaced by a temporary one. Up to this point, however, the most controversial project has been the bobsled slope planned to replace the existing one built in the 1950s for the 1956 Cortina Winter Olympics. This project would require razing three hectares of forest and constructing new infrastructure such as a pumping station to divert water from the Boite river to cover the slope in ice. Since 2021, this bobsled development plan has been fiercely debated and contested by environmentalists, citizens, mountain associations, and some local inhabitants, not only because of the environmental impact it would have on the mountain slope but also and especially because of its cost, subsequent maintenance requirements, and the limited time available for construction. The projected budget reached approximately EUR 120 million and no construction companies applied for the first two tenders. Since fall 2023, therefore, Olympics and political institutions have been discussing the possibility of moving bob activities to an existing structure in Kitzbuhel, Austria. Recently, a third tender launched to keep the bobsled slope in Cortina secured the interest of one company and its construction was approved in February 2024; Since fall 2024 work has been in progress, three hectares of forest were razed and there are still uncertainties with regard to the project completion.

The initiatives promoted by Olympics institutions together with the Veneto provincial government and Cortina city council thus clash dramatically with the Olympics nomination dossier rhetoric highlighting the use of existing structures and special attention to sustainability. Indeed, the president of the Parco Naturale Dolomiti d'Ampezzo, managed by the Regole d'Ampezzo, argued that these initiatives – especially the bobsled slope – are controversial specifically in terms of their environmental-economic impact and the way they exploit mountain nature. A key piece of this picture is the fact that, in February 2022, a national governmental decree placed Milano Cortina 2026 and its initiatives under a special administrative regime to speed up bureaucratic procedures related to infrastructural development. The specific conditions of this regime lessen environmental regulation constraints, especially by both postponing and hastening strategic environmental impact assessments (VAS). The VAS in this case was finally conducted in spring 2024, three months after work had already been initiated to build the bobsled slope.[4] Indeed, the former president of the association Mountain Wilderness criticized this special administrative regime and stated that, without the regime, the planning phase of the VAS process would have two elements: first, a socio-environmental impact assessment of all Olympics activities and infrastructural development plans; and second, involving local stakeholders in decision-making processes and supporting participatory procedures designed to open up an inclusive space of debate about projects. A member of the Cortina city council argued that postponing the VAS constituted a deep betrayal of the organizing institutions' avowed commitment to sustainability, destroyed the credibility of their narratives, and severely limited the transparency of Milano Cortina 2026 organizational procedures. Indeed, the special administration regime revealed the controversies surrounding Olympics politics and the uneven power relations involved in their decision-making processes. The president and three interviewed members of the Comitato Civico Cortina stated that the Cortina mayor was in some way forced by the Olympics commissioner and regional and national political institutions to approve development initiatives by avoiding discussion in the city council and limiting the information shared with citizens. The former president of the Italian Alpine Club highlighted this lack of transparency by arguing that Cortina locals are not aware of either the decisions being made by local elected officials or the potential socio-environmental impacts of planned infrastructure that could affect the valley. This situation had thus enabled a top-down process of controversial mountain exploitation concealed by public discourse about a collective dream of sustainability, futures, and legacy, as highlighted by newspaper articles and posters hung up around Cortina.

The dream of mountain nature futures that we see in this case, produced by powerful transcalar actors, thus excludes the majority of Ampezzo community members and associations. Furthermore, this mountain nature futures vision – aimed at reproducing the commodification of nature through mega-events – significantly downplays the destructive nature of infrastructure development, especially the impact of climate change on winter sports (Casanova, 2022; Silk, 2014). Indeed, a member of the Regole d'Ampezzo argued that Olympics infrastructural development initiatives have contributed to exacerbating existing conflictual relations with the local Cortina government around mountain natures perspectives. In order to understand the politics, visions, and discourse of Dolomites mountain futures, and especially their contested nature, it is important to also consider another project, the *Carosello Dolomitico*. This initiative has been promoted since 2020 by the Veneto regional president with the support of multiple mountain lift companies in the wake of the 2021 World Ski Championships and Milano Cortina 2026 development process. The *Carosello* is conceived as an intra-valley network of ski-lift connections linking Cortina with other villages and valleys of the Dolomites, especially the Civetta and famous Sella Ronda ski resorts. The promoters' idea is to connect and extend the Dolomites' ski area to 1300 km by building long lifts and gondolas, up to 7 km in length, and developing new ski slopes. According to the Veneto government and mountain lift companies such as ISTA and Rete Dolomiti, the *Carosello* – to be launched in conjunction with the 2026 Olympics – will allow visitors to reach the Alps' most spectacular valleys without ever taking off their skis, thus becoming the largest ski resort in the world.

The Veneto provincial government, and especially its president Luca Zaia, has emphasized the project's key role in enabling visitors to experience the beauty of Dolomites nature by connecting Veneto valleys with those of the neighboring Trentino-Alto Adige province; in particular, however, it would grant a socio-economic boost to smaller ski resorts and surrounding villages while also promoting sustainable development and green tourism. Indeed, from the perspective of the provincial government, the *Carosello* is intended to prevent the marginalization and depopulation of mountain villages and, at the same time, foster sustainable mobility, especially in the summer, by reducing intra-valley car travel. Moreover, the actors pushing for this project have stated that it represents a mountain protection initiative for sustainable development in that it aims to limit social marginalization and private transportation pollution.

Reflecting on the *Carosello Dolomitico* initiative combined with the Milano Cortina 2026 organizational process and the discourses of promoting actors, we see a shared logic of mountain nature exploitation and branding through infrastructure. Olympics authorities, political institutions, and mountain lift companies have continued to adhere to a mountain nature commodification perspective rooted in the modernist capitalist approach characterizing mountain development since the 1970s, but now promoted through a sustainability-fostering discourse despite the significant contemporary effects of climate change. Indeed, the director of Funivie Arabba emphasized the role new technologies play in mitigating climate change effects, especially in relation to producing artificial snow, and stated that these dynamics will not hamper the sustainable development of the mountain in the future. And yet conversations with representatives of Comitato Civico Cortina, Regole d'Ampezzo, and the director of the Parco Naturale Dolomiti d'Ampezzo show that these initiatives – particularly the *Carosello* – and their associated discourses are highly contested. Critics point to the effects of the eco-climate crisis, the environmental impact of projects, and the massive use of public funds for private profit. Indeed, the *Milano Cortina 2026* organizational process and *Carosello Dolomitico* promotional campaign, combined with the increasingly serious effects of climate change in the Dolomites manifesting since 2020 in snow cover reduction, records heat levels and drought, have acted as a catalyst driving people to question the logic of nature-exploitation underpinning these initiatives. For instance, the *Carosello* in particular has been contested by the residents of Fodom valley, located between Cortina and Arabba. They formed a committee to denounce the project's severe environmental impact on forests and high-altitude meadows, many of which are included in Natura 2000, and stress the importance of preserving the area's historical and cultural heritage from new infrastructural construction. The mayor of Colle Santa Lucia, located near the Civetta ski resort, added that, whereas in 2020 the majority of residents supported the intra-valley connection, today the community is divided due to the initiative's controversial socio-environmental and economic aspects. Collective reflection on the contestable aspects of the mainstream logic of mountain commodification through infrastructures and discourses is thus giving rise to a counternarrative aimed at moving beyond this political perspective. Over the last few years, this alternative vision has been taking hold thanks to the activities of an informal network of sociocultural and environmentalist associations including Mountain Wilderness, Legambiente, WWF, Libera, and Italia Nostra as well as mountain guides, local city council members, journalists and academics, other individuals, and local valley associations such as the FODOM committee.

According to this heterogenous counternarrative, Cortina and Dolomites mountain commodification initiatives and infrastructural development projects are not sustainable for three main reasons. The first is their environmental impact on nature and lack of effective impact assessments; the second is their questionable use of public funds to support private business and profit-making. Third, they are characterized by top-down politics and a lack of transparency and community involvement. Recently, in January 2024, these claims have been supported by the launch of an online magazine called *L'Altra Montagna*, issued by the *Il Dolomiti* newspaper in collaboration with mountain experts and scientists. The goal of the magazine is to disseminate information about interactions between the effects of the eco-climate crisis and the socio-environmental challenges facing mountain communities, especially by supporting a radical change in visions of mountain futures. The focus is not restricted to the Dolomites and Italian Alps but addresses mountain environments in general. *L'Altra Montagna* thus supports a social and political transformation aimed at reconfiguring mountain society-nature interplays in the eco-climate crisis in a more just direction.

Conclusion

This chapter contributes to the discussion on Italian social natures and their commodification in the contemporary eco-climate crisis through a focus on mountain natures in the Dolomites. It has analyzed the politics, visions, and discourses of the Milano Cortina 2026 Winter Olympics by examining infrastructural initiatives promoted by governance actors. In terms of theoretical and methodological framework, the research combined a political ecology perspective based on exploring contested mountain futures with ethnographic research in the Ampezzo valley. The evidence shows that the assignment of Milano Cortina 2026 has been key in the process of revamping the image of Ampezzo valley mountain nature by reproducing a logic of mountain commodification. Indeed, this process – deeply rooted in the conservative mountain development vision that has characterized the last few decades – has involved two main axes. On one hand, exploiting mountains through new infrastructural projects with significant environmental impact; on the other hand, 'branding' mountains by emphasizing discourses of sustainability, green mobility, inclusiveness, and legacy that reproduce the exploitation and commodification logics identified by Castree (2014) and Büscher and Fletcher (2020), among others. This process of mountain nature commodification, characterized by a depoliticized and naturalized valorization and an accumulation-by-sustainability vision, does not seem to take seriously the present and future effects of the eco-climate crisis, especially on winter sports and summer water availability.

Furthermore, the on-going mountain commodification process is characterized by public capital flow on multiple levels to support private infrastructures and profit. At the same time, it involves top-down politics driven by external powerful actors and uneven power relations that aim to limit the involvement of mountain communities and their potential contestation. In view of these controversial aspects, this logic has being increasingly challenged by a counter-narrative pointing out the unsustainability of this mountain commodification process in the context of eco-climate crisis and calling for a radical change in mountain development politics and society-nature relations. This perspective of change stresses the need to adapt socio-economically to the eco-climate crisis by rethinking notions of limits and value in the direction of more just mountain futures. Embracing the insights of Castree (2014) and Loftus (2017), therefore, even though the social production of mountain nature in the Dolomites is still dominated by an accumulation-by-sustainability vision, it does embody a growing counternarrative of change that generates ever-growing contestation and might lead toward more just futures. This counternarrative is focused not on boycotting sport mega-events such as the Olympics in a short-term perspective, but at questioning these processes of nature commodification and social exclusion and fostering a radical change in mountain politics and socio-economic development to rebalance socio-environmental relations in a long-term perspective. As also reported by *L'Altra Montagna*, Dematteis and Nardelli (2022), and Albino Ferrari (2023) among others, such rethinking concerns not only the Dolomites but also other Italian mountain natures undergoing transformation, from the Central-Western Alps with the cases of Gavia Pass and Vallone Cime Bianche, to the Northern Apennines with the case of Corno alle Scale.

Ethnographic research, combined with a political ecology perspective, offers a deep understanding of discourses, governance actors' visions, and related power relations in these contexts. Also and more importantly, taking on Kothari et al.'s (2019) and Benjaminsen and Svarstad's (2021) arguments, it highlights the importance of supporting representatives of mountain communities in pushing for a radical change in mountain socio-environmental futures. Indeed, this combined theoretical and methodological perspective applied to mountain futures in the eco-climate crisis contributes significantly to the geographical understanding of Italian socio-natures by emphasizing their power relations and conflictual dimensions. At the same time, it strengthens the relationship between political ecology and mountain geography and advances discussions about the need to reconfigure mountain futures and foster counternarratives about them.

Notes

1 The website of the Fondazione Dolomiti UNESCO is https://www.dolomitiunesco.info/.
2 https://www.glaciologia.it/i-ghiacciai-italiani/
3 https://milanocortina2026.olympics.com/it
4 As an environmental regulation initiative, Valutazione Ambientale Strategica (VAS) has been used since 2001 by the European Union to evaluate and monitor the environmental impact of infrastructural development plans and programs involving multiple, diverse infrastructural elements. The aim is to preserve the environment

and ensure the sustainability of planned infrastructure. The VAS is usually conducted by the state administration such as provincial or district governments as part of the planning phase of projects and programs. The environmental impact assessment tool Valutazione d'Impatto Ambientale (VIA) is instead focused on gauging the environmental impact of individual pieces of infrastructure and is conducted by the same authorities during the design phase. Another kind of assessment, Valutazione d'Incidenza (VINCA) or evaluation of incidence, must be conducted for any kind of infrastructural development that might have a direct or indirect impact on flora and fauna habitats included in the European Union Natura 2000 Network (sources: ISPRA and Regione Veneto websites).

References

Albino Ferrari, M. (2023). *Assalto alle Alpi*. Einaudi.

Benjaminsen, T. A., & Svarstad, H. (2021). *Political ecology: A critical engagement with global environmental issues*. Palgrave Macmillan.

Bonati, S., Tononi, M., & Zanolin, G. (2021). Social nature geographies. Le geografie e l'approccio sociale alla natura. *Rivista Geografica Italiana*, *CXXVIII*(2), 5–20. https://doi.org/10.3280/rgioa2-2021oa12029

Bryant, R. L. (2017). *The international handbook of political ecology*. Edward Elgar.

Büscher, B., & Fletcher, R. (2020). *The conservation revolution: Radical ideas for saving nature in the Anthropocene*. Verso.

Casanova, L. (2020). *Avere cura della montagna. L'Italia si salva dalla cima*. Altreconomia.

Casanova, L. (2022). *Milano Cortina 2026 – Ombre sulla Neve*. Altreconomia.

Castree, N. (2003). Environmental issues: Relational ontologies and hybrid politics. *Progress in Human Geography*, *27*(2), 321–334. https://doi.org/10.1191/0309132503ph422pr

Castree, N. (2014). *Making sense of nature: Knowledge, politics and democracy*. Routledge.

Debarbieux, B., & Price, M.F. (2008). Representing mountains: From local and national to global common good. *Geopolitics*, *13*(1), 148–168. https://doi.org/10.1080/14650040701783375

Dematteis, M., & Nardelli, M. (2022). *Inverno Liquido*. DeriveApprodi.

Ernstson, H., & Swyngedouw, E. (2019). Bringing back the political: Egalitarian acting, performative theory. In H. Ernstson, & E. Swyngedouw (Eds.), *Urban political ecology in the anthropo-obscene: Interruptions and possibilities* (pp. 255–267). Routledge.

Ferrario, V., & Marzo, M. (2020). *La montagna che produce*. Mimesis.

Fonstad, M.A. (2017). Mountains: A special issue. *Annals of the American Association of Geographers*, *107*(2), 235–237. https://doi.org/10.1080/24694452.2016.1260898

IPCC Climate Change. (2022). *Impacts, adaptation and vulnerability. Contribution of working group II to the sixth assessment report of the intergovernmental panel on climate change*. Cambridge University Press.

Kothari, A., Salleh, A., Escobar, A., Demaria, F., & Acosta, A. (2019). *Pluriverse: A post-development dictionary*. Tulika Books.

Loftus, A. (2017). Political ecology I: Where is political ecology? *Progress in Human Geography*, *43*(1), 172–182. https://doi.org/10.1177/0309132517734338

Martini, A., Colombino, A., & Zinzani, A. (2022). Metodi e metodologie per la ricerca sul campo in geografia. In C. Minca (Ed.), *Appunti di Geografia* (pp. 443–465). Cedam.

Perreault, T., Bridge, G., & McCarthy, J. (2015). *The Routledge handbook of political ecology*. Routledge Earthscan.

Silk, M. (2014). Neoliberalism and sports mega-events. In J. Grix (Ed.), *Leveraging legacies from sports mega-events: Concepts and cases*. Palgrave Pivot. https://doi.org/10.1057/9781137371188_5

Smith, N. (1984). *Uneven development: Nature, capital and the production of space*. Basil Blackwell.
Stoddart, M. C. J. (2013). *Making meaning out of mountains: The political ecology of skiing*. UBC Press.
Varotto, M. (2020). *Montagne di mezzo. Una nuova geografia*. Einaudi.
Voicu, S., & Vasile, M. (2022). Grabbing the commons: Forest rights, capital and legal struggle in the Carpathian Mountains. *Political Geography*, *98*, 1–11.
Zinzani, A. (2020). L'ecologia politica come campo di riconcettualizzazione socioambientale: governance, conflitto e produzione di spazi politici. *Geography Notebooks*, *3*(2), 33–50. https://doi.org/10.7358/gn-2020-002-zinz
Zinzani, A. (2023). The contested environmental futures of the Dolomites: A political ecology of mountains. *Geographica Helvetica*, *78*(2), 295–307. https://doi.org/10.5194/gh-78-295-2023

8 The co(a)sts of beauty

From the construction of the tourist imaginary of attractiveness to increased exposure to risk*

Sara Bonati, Eleonora Gioia, and Eleonora Guadagno

Introduction

The preservation of the "beautiful Italian coast" has long been a priority of the national, regional, and local branding processes, where coasts occupy an important space in the tourist promotion of the "Bel Paese", as Italy was already called in the 13th to 14th centuries by Dante Alighieri in the *Divina Commedia* and Francesco Petrarca in the *Canzoniere*. Thus, by "beautification" we mean the processes of imaginary construction and landscape transformation that have contributed, among others, to the idea of Italy as "the beautiful country", transforming it into one of the most famous worldwide tourist destinations. In 2022, 26.9% of Italy's foreign tourists were attracted by the coasts, thus representing the second reason to visit Italy after cultural tourism, producing an income of around €76 billion (28.8% of the total holiday spending; Banca d'Italia, 2023). Therefore, tourism on Italian coasts, as more broadly in the Mediterranean, is considered a major source of growth.

As a consequence of their beautification since the 1960s, the Italian coasts have experienced a concentration of investments in massive land use processes that have been translated into resource exploitation, hyper-urbanisation, and hyper-touristification, all related to the concept of "coastalisation". Indeed, "coastalisation" here identifies the processes of population/activity concentration and urban sprawl that take place in coastal zones, especially intensified by tourism urbanisation (among others, Lagarias & Stratigea, 2023). The process is not without consequences for the territories. On the one hand, this diverts attention from the hinterland, favouring a dichotomous relationship between coasts and internal areas (often abandoned and thus more vulnerable to hydrogeological phenomena); on the other hand, it produces negative consequences regarding the coastline's vulnerability, impacting the resilience of land and marine ecosystems (Lagarias & Stratigea, 2023) and humans' ability to enjoy these landscapes.

* Although the chapter is the result of a joint collaboration among the authors, Sara Bonati wrote paragraph 1, Eleonora Gioia wrote paragraph 2, and Eleonora Guadagno wrote paragraph 3, while the conclusions were collectively authored by all three.

DOI: 10.4324/9781003465164-11

In addition, today, climate change is an increasing threat to the Italian coasts, producing, among others, sea level rise, coastal flooding, increased extreme weather hazards, and coastal erosion. Nowadays, 46% of the sandy coasts in Italy suffer from coastal erosion, with the highest values in Abruzzo, Sicily, and Calabria (up to 60%); since 1970, the coastline stretches subject to erosion have tripled (Legambiente, 2022). The loss of coastline is also intensified by human activity, which has produced a loss of 5 km/year in the last 20 years, corresponding to 100 km (Munafò, 2023).

Despite this increasing exposure of most of the Italian coasts to risk and the national and international calls to adopt a socioecological lens on disasters, the approach to disaster risk reduction has not changed in Italy in recent decades, continuing to favour a technocratic interpretation (Guadagno & Grasso, 2022) unable to adequately discuss and address the negative consequences produced by beautification and coastalisation processes. This derives mainly from a "naturalising" narrative of disasters, in which the role of human activity in the construction of risks is veiled, and responsibility is diverted towards "Nature" or the "Climate" as part of a process of personifying the two entities (Pelling, 2001).

Accordingly, this chapter aims to identify the processes of nature exploitation that are the basis of beautification and coastalisation and how they are related. This contributes to understanding how territorial differences and the commodification of nature are built by the two processes, such as how they are the basis of risk production (Cutter & Finch, 2008). Within this scope, the idea of "sand urbanisation" suggested by Dawson (2021) is here adapted and extended. If Dawson uses the expression to define the city as a "geological construction" or "sedimentary system", in this paper, the idea of sand is expanded considering how sand and more broadly nature participate in building tourist imaginaries and in favouring coastal urbanisation. Thus, we try to "extract" the role of sand and nature in building the coast and becoming urbanised. Two aspects are mainly considered here: on the one hand, how coastal cities feed off sand/nature to produce urban imaginaries and become attractive, according to beautification; and, on the other hand, how sand/nature is part of material city building and coastalisation, becoming an object of consumption for extending the borders of the human settlements and the *topos* of conflict among actors compelling for the same scarce resource city.

The analysis is based on two areas of Marche and Campania, among the most exposed Italian regions to the coastal effects of environmental degradation and the consequences of climate change (ISPRA, 2024). Furthermore, they represent two different examples of Italian coastal exploitation. In the wake of a research project initiated by the authors in various coastal contexts (Celata & Gioia, 2024; Casareale et al., 2023; Donadio et al., 2023; Gioia & Guadagno, 2024; Guadagno & Grasso, 2022), the two cases—Senigallia with its "Velvet Beach", and Ischia, the "Green Island"—are paradigmatic. On the one hand, they represent an image of "beautiful" coastal tourism par excellence, and on the other hand, they are areas that, due to the processes of nature commodification and their intimate connection to tourist monoculture, have become particularly exposed to underlying risks and amplified their vulnerability.

Accordingly, this chapter is based on secondary data and scientific literature to rebuild and redraw the connections between the different processes and challenge the concepts used in the analysis, to understand their potential to promote a socioecological perspective and better understand coastal futures.

Senigallia: The city of the "Velvet Beach"

Senigallia, a municipality located on the central-northern coast of the Marche region, approximately 30 km north of Ancona, plays a primary role in the regional panorama due to its tourist vocation. With a population of about 44,000 inhabitants, the municipality covers an area of approximately 120 km^2 and is situated at the mouth of the river Misa. The territory is mainly flat due to its location in an alluvial plain, but it is surrounded by hills that slope down towards the sea from the highest point of approximately 200 m above sea level.

The first hotel in Senigallia was built by the new ruling class that assumed city governance following the establishment of the Kingdom of Italy in 1861 (Pupazzoni, 1975). This was accompanied by state-sponsored interventions to create the railway and develop the port, to make Senigallia a hub of commercial activity and reduce its isolation, partly due to the orographic conditions and the declining longevity of the summer free-trade fair (Cassani, 2020). However, as the project to transform the area into an industrial hub failed, the ruling class opted for coastal tourism.

At the turn of the 19th and 20th centuries, Senigallia already had important tourist value, which increased in the following years: symbols of this phenomenon were the creation of the Joint-Stock Company of the Sea Baths Establishment (in Italian, *Società per azioni dello Stabilimento per i Bagni Marini*) (Monti Guarnieri, 1961) and the *Rotonda a Mare*, once a stilt house located in front of the *Bagni* establishment for its use and rebuilt in its current position with reinforced concrete in 1933 after the destructive 1930 earthquake (Santoni, 2017). Gradually, the municipality became involved in the management of the baths, eventually assuming control of all their activities. This led to the formulation of a plan for the expansion and development of tourism (Pupazzoni, 1975). The trade fair season became the bathing season, with the accompanying facilities and services.

The pivotal role that the city assumed in the tourism sector was formally acknowledged in 1928, when Senigallia and Cortina d'Ampezzo were designated as the inaugural locations of an Autonomous Healthcare and Accommodation Company in Italy. This entity's multifaceted responsibilities encompassed a range of services, including reception and tourist information, event promotion, lifeguard management, beach cleaning, and the collection of the tourist tax (Goffi, 2010). Since then, the transformation of the coast into a tourist space has resulted from a historical process of beautification aimed at enhancing the coastline. This has involved the active promotion of natural beauties, the creation of an idyllic image, emphasising its landscape and cultural peculiarities, and—of course—the design of tourist infrastructures. For decades, Senigallia has been the subject of advertising campaigns that have defined it as the city of the "Velvet Beach" to attract the tourist's imagination, recalling the sensation you get when stepping on soft sand under your feet, just like velvet. The expression "Senigallia, Velvet Beach" was coined in the mid-1930s by a journalist from *La Stampa* of Turin, Mario Carafoli, born in a neighbouring city, at the request of the then head of the Senigallia tourist company (Santoni, 2017). The slogan achieved considerable success and survived to the present as a city brand, to the point of even becoming a parfum.[1]

Throughout the 1950s and 1960s' "economic miracle", the city went through a strong period of growth and tourism as a sign of recovery from the passage of the front of the First and Second World Wars. In that period, Senigallia experienced a true explosion of seaside tourism, sealed by a famous song by Fred Bongusto from 1964 dedicated to the town's *Rotonda a Mare*. Yet, such an established brand did not escape the seaside product's phase of decline at the end of the century. Indeed, following the economic boom of the 1960s and 1970s, from the mid-1980s onwards, there are indications that tourist flows in Senigallia, as in other Italian seaside destinations, have been in decline (Goffi, 2010). Signals of recovery followed in the 1990s and 2000s. The municipal public administration has thus committed to achieving the greatest possible attractiveness with some events that nowadays have national and international fame, namely the *CaterRaduno* and the Summer Jamboree. These events have contributed to the town's relaunch (Goffi, 2010). The latter has become one of the three most significant festivals of 1940s–1950s music in the world, characterised by a sense of revival. This encompasses a return to the 1950s, the post-war period, which also represented a rebirth for Senigallia. Moreover, since 1997, the city has been continuously awarded the "Blue Flag", an eco-label granted by the Foundation for Environmental Education (FEE) to seaside tourist destinations that meet sustainable sea and land management criteria.

Today, Senigallia is one of the most important seaside municipalities in the Marche region regarding accommodation and the number of beds, arrivals, and visitors (Goffi, 2010). According to the Marche Tourism Observatory, in 2022, Senigallia was the fourth municipality of the region in terms of number of travellers, with 175,189 arrivals, and the first in terms of overnight stays, with 876,589 total nights (Marche Tourism Observatory, n.d.-a). It is classified by ISTAT as a municipality with a "very high tourist density" (ISTAT, 2022). Indeed, between 2002 and 2022, the number of tourist accommodation establishments increased from 142 to 193 (ISTAT, n.d.). This corresponds to an accommodation density (i.e., the number of total accommodation facilities per km^2) increase of 58% in 2016–2021 alone (Marche Tourism Observatory, n.d.-b). However, over the years, this effort has resulted in the coastal area's intense development. According to a report by the Italian Higher Institute for Environmental Protection and Research (ISPRA), the Municipality of Senigallia is experiencing land consumption of approximately 11%, compared to an average of about 7% for the regional and national territories (Munafò, 2023). In 2022, the land consumption within 300 m of the shoreline was greater than 57%, and the land consumption between 300 and 1,000 m was nearly 40%. It is notable that between 2006 and 2022, an increasing trend was observed in the 0–300 m and 300–1,000 m strips, with rates of 1.3% and 1.7%, respectively. This trend, likely linked to the increasing investment in tourist facilities and services, is in stark contrast with the city's ongoing issues with coastal erosion. In fact, according to ISPRA (2023), of the 13.4 km of low natural coastline, between 2006 and 2020, 28.8% has undergone significant landward movement, with a displacement of more than 5 m (Figure 8.1). This is in line with the regional coastal erosion value of 27.9%, which is expected to worsen significantly due to future climate change scenarios (Celata & Gioia, 2024).

In this sense, Senigallia is facing a critical paradox: while the development and expansion of tourism infrastructure are consuming more coastal land, the shoreline itself is retreating due to erosion. As more land is developed for hotels, restaurants, and other tourist facilities, the natural coastline diminishes, reducing the space available for recreational activities. Moreover, the intensifying coastal erosion driven by climate change exacerbates this issue, as rising sea levels and increased storm intensity lead to more substantial and frequent loss of beachfront. In addition to these factors, the city's location at the mouth of

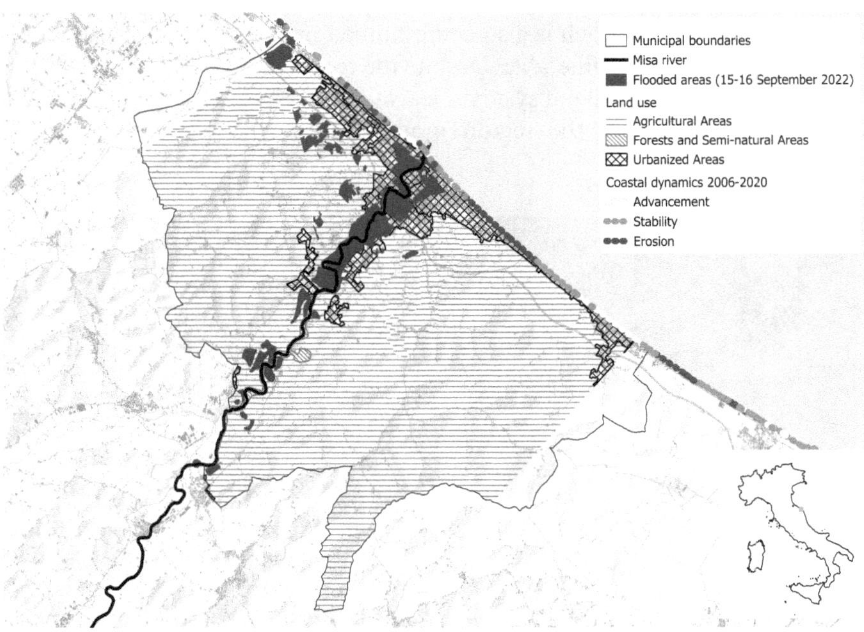

Figure 8.1 Map of Senigallia: (i) inundated areas after the flooding of the river Misa on 15–16 September 2022, (ii) land use, and (iii) shoreline dynamics between 2006 and 2020 (elaboration by the authors, based on the Copernicus Emergency Management Service—Mapping https://emergency.copernicus.eu/mapping/list-of-components/EMSR634 and ISPRA https://sinacloud.isprambiente.it/portal/apps/sites/#/coste/pages/dati).

the river Misa results in encroachment upon the land from the river delta and periodic inundations. Senigallia faces a primary risk of fluvial and coastal flooding. Historically, significant flooding events in 1940, 1955, 1976, 2011, 2014, and 2022 resulted in numerous casualties and extensive damages (Figure 8.1) (Gioia, 2023; Santoni & Morici, 2021). Such conditions can be attributed to two primary factors: first, the geomorphology of the river itself, which is characterised by a torrential nature; and second, the shrinkage of the river mouth due to the process of urbanisation (observable in Figure 8.2) which, over the years, has led to the sprawl of commercial and port activities, residential areas, and tourist infrastructures. The concentration of people and resources along the city centre and the coast has thus increased anthropogenic pressure and rendered these areas more vulnerable and more exposed to extreme events such as storms, floods, and coastal erosion, consequently exacerbating the flooding risk.

In this sense, the case of Senigallia exemplifies the paradigmatic nature of the Italian situation, with responses to coastal issues based (for much of their history) on localised structural protection interventions focused on specific sectors. On the one hand, tourism continues to increase, drawn by the allure of the "Velvet Beach" brand; on the other hand, the beach itself is diminishing due to both natural causes and the effects of coastalisation and climate change. This reduction of the beach is also compounded by increasing floods, making Senigallia particularly vulnerable. Despite the recurrent threats, Senigallia currently lacks adequate resilient systems, leaving the city with the highest number of tourist presences in the region more and more susceptible to future

Figure 8.2 Coastal area of Senigallia in 2023 (Google Earth); the shaded polygon indicates the extent of the inhabited centre in 1889, based on the authors' elaboration of data from Santoni & Morici (2021).

flood-related disasters. This unsustainable path could ultimately diminish Senigallia's attractiveness as a tourist destination, underscoring the need for a balanced approach to development that considers long-term environmental impacts.

Ischia, or the vulnerable "Green Island": too beautiful to be true

In spite of its narrative deeply intertwined with a tourist ideal, the Island of Ischia, one of the "pearls" of the Campanian Archipelago (Ruocco, 1988), has recently served as the backdrop for tragic calamities connected to hydrogeological instability and arson, consistently highlighting a sustained assault on its territory attributable to escalating instances of building abuse, enchanted by its beauty and the narrative surrounding it (Figure 8.3).

Situated approximately 30 km from Naples and connected via ferries and hydrofoils to the mainland, the island spans an area of 46 km^2. It is recognised as the emerged portion of a volcano within the Flegreo volcanic district (De Alteriis & Violante, 2009). The morphology of the island and the characteristic colour of the tuff that forms its backbone have earned it the appellation of the "Green Island", a designation also linked to its lush vegetation. Ischia is an important centre of biodiversity in the Mediterranean due to its microhabitat, home to a considerable number of rare and endemic species, some of which are highly endangered (Monti et al., 2008; Sibilio et al., 2015). This quality possesses a "fascination truly inseparable from the landscape", as noted over a century ago by Sybil Fitzgerald, a cultured American traveller who vacationed on the island (Fitzgerald & Fitzgerald, 2008).

Ischia has an average altitude of 150 m above sea level, featuring a notable peak, Mount Epomeo, reaching 787 m. The landscape includes hilly reliefs from minor eruptive centres and high cliffs along the southern coast,

Figure 8.3 Barano d'Ischia (NA), Maronti beach. Debris fall. Photo by the authors (2023).

constituting 70% of the island's perimeter, while the northern region has low, sandy coastlines. Geologists, including Breislack, Mercalli, and Rittman, have studied the island's geomorphological characteristics since 1789. Ischia experiences robust seismicity unrelated to volcanic activity, making it susceptible to catastrophic events that have multiplied over the past 50 years, primarily due to illegal construction and inadequate water management (Celico et al., 1999; Esposito & Matano, 2023). The volcanic nature of the soils, coupled with factors like intense rainfall and seismic activities, contributes to recurrent landslide phenomena documented since the Middle Ages (Mele & Del Prete, 1998). Anthropogenic factors, particularly the abandonment of traditional sectors (namely agriculture and silviculture) and tourism-related urbanisation, also increase the island's vulnerability to such events. The hazardous conditions, exacerbated by ongoing changes in meteoclimatic patterns, highlight the complex interactions between humans and the environment in this fragile island ecosystem.

Despite its inherent hazards, the island has been inhabited for millennia, primarily due to its fertility, favourable climate, and strategic location in the Mediterranean basin. Its peak occurred under the Spanish House of Avalos and later during the reign of King Ferdinand II, who in 1854 directed the removal of the mouth of the volcanic lake, transforming it into the island's primary port. This development established Ischia as a hub for the export of wine, other agricultural products, and fish, accelerating the island's reputation, economy, and subsequent urbanisation (Mazzetti, 2006).

However, since the 1950s, due to its reputation as the "Green Island", there has been a substantial surge in construction activities closely associated with uncontrolled tourism development, drawn by its beauty. This began in niche markets and later expanded to encompass mass tourism in coastal areas overall. This trend followed the island's rediscovery by publisher Angelo Rizzoli, who transformed it into an internationally renowned spa resort. He consolidated his hotel empire in Ischia by acquiring the Regina Isabella Baths and the Radium Santa Restituta Baths, creating the Regina Isabella Hotel—a flagship on the island. He expanded his holdings with the acquisition and renovation of Hotel Terme Manzi and the construction of Hotel Reginella, complete with a cinema and a theatre.

During the 1950s, Rizzoli utilised his publishing and film resources to promote Ischia's image, beauty, and interests. Films like *The Awakening* (*Suor Letizia*, 1956) and *Holidays in Ischia* (*Vacanze ad Ischia*, 1957) played a role in a territorial marketing strategy that emphasised the island's beauty, its lush, almost "wild" landscapes, and breathtaking scenery aimed at capturing the attention of the international jet set, and the island is currently the venue for a renowned international film festival. Suffice it to say that an internationally renowned couture house unveiled its 2022 collection, evocatively named "Local Color" after Truman Capote's collection of short stories that includes a novella recounting the author's four months spent on the island and evoking the splendour of the era when high society tourist routes had Ischia as their destination.

However, Rizzoli's death in 1970 marked the gradual decline of élite tourism: Ischia evolved into a mass tourist resort, attracting the German market and the Italian middle class (Delizia, 1987).

Presently, the island is divided into six municipalities, collectively accommodating around 62,000 residents. This number experiences exponential growth during the tourist season, when, for example, the island struggles to maintain adequate water service and must rely on water from cisterns. Additionally, in the summer, the seawater is filled with mucilage and foam (Gambi, 2014)[2]: this depletion in quality and quantity is also exacerbated by the impacts of climate change, particularly due to drought events and salinisation. Geared towards popular and wellness tourism, drawing visitors primarily from the Naples metropolitan area, Ischia caters to a local and regional audience seeking relaxation (Monti, 2006). Furthermore, it positions itself as an attractive destination for spa packages (also thanks to reimbursements from the national health care system, as regulated in Italy by Act 323/2000 on the reorganisation of the spa industry), particularly appealing to tourists from Eastern Europe and Germany.

Nowadays, all six municipalities are classified as extremely "densely frequented by tourism", according to ISTAT's 2022 classification. Over the past four decades, the average degree of urbanisation has surged almost sixfold, rising from 7% to 40% of the total land area (ISPRA, 2022). This situation of over-exploitation of the littoral also emerges from the CORINE land use cover data viewer, especially when comparing built-up areas over the last few decades or even using the Google Earth time-lapse engine. Accordingly, from ISPRA data, it emerges that the island has seen land consumption of 15 hectares during the last 15 years: on average, 10,000 m^2 per year of new construction (2022).

This phenomenon has played a substantial role in irreversibly compromising the island's hydrogeological structure over time (Gugg, 2018). The latter can be attributed to significant urbanisation and tourism activities on the island, particularly along coastal regions, and this includes the gradual abandonment of agricultural activities and rural–urban migrations to the major centres of the island or the mainland (as documented by Alberico & Petrosino, 2014; Buchner Niola, 1965; Di Martire et al., 2012). These activities have caused extensive alterations to the hydrographic network, both visible on the surface and underground. This has led to increased spatial and temporal recurrence of instability phenomena, as Del Prete and Mele (2006) and Budetta et al. (2013) noted.

What will be left of Ischia over time if the exploitation process continues at these rates? Is the island already at its saturation point? What is detected beyond the predictions is that these phenomena significantly impact the island's susceptibility, contributing to heightened hazards such as landslides and hydraulic events, thereby adversely affecting the overall quality and resilience of the island's landscape (see also, and in consideration of Capri: Russo Krauss, 2023). Paradoxically, the 2022 *#ThisIsIschia* campaign, supported by the Ministry of Tourism to rejuvenate tourism after an impactful earthquake in 2017 and a destructive landslide in 2022, primarily affecting unauthorised or tolerated dwellings (Sisma ed Emergenza a Ischia, 2023), despite the process of beautification, appears to inadvertently perpetuate detrimental tourism practices that contribute to an increase in squatting; considerably, the promotional campaign was also disseminated in Paris, where the hashtag and an image portraying the Aragonese Castle are prominently featured in a prestigious showcase at the Place de l'Opera.

In this way, and quite ironically, Ischia becomes a simulacrum of its own tourist imaginary. On the one hand, this imaginary has wrapped it in cellophane and adorned it through marketing processes, thus leading to increasingly extreme monocultural exploitation. On the other hand, it has impoverished the island of its resources, viewing them merely as "territorial capital", thereby weakening it environmentally. The challenge becomes more apparent because, despite marketing efforts portraying "Ischia" as a unified entity for greater recognition, even at the international level, the island's fragility is heightened by administrative fragmentation which hampers the development of a comprehensive perspective on multi-hazard and cross-border situations. As a result, insufficient land management practices and the excessive and neoliberal exploitation of the island's resources, combined with an aggressive push for tourism through marketing, highlight a political disregard for local economic and environmental intricacies. In practical terms, this oversight leads to heightened exposure to risks and increased social vulnerability on the island (Donadio et al., 2023).

Conclusion

This chapter has discussed the role of nature/sand consumption in producing tourism-related urban sprawl and activating or interacting with risk dynamics. To this purpose, the processes of beautification and coastalisation have been investigated. In particular, an example of sand urbanisation (Senigallia) and one of nature urbanisation (Ischia) have been analysed.

In particular, Senigallia has shown how sand has contributed to building performative imaginaries of the coast and to the process of beautification that has activated the processes of hyper-urbanisation and hyper-touristification. In this case, "coastalisation" is, first of all, a process of coastal imagination that has commodified the beach/sand, particularly for tourism purposes.

The second case, Ischia, concerns how the nature consumption activated by urban expansion has triggered risk construction mechanisms, making the coasts fragile and vulnerable. Furthermore, increasing risks associated with climate change are taking place on the island, intensifying the competition between residents and tourists in the use of local resources. This is especially the case of water but also land. Thus, in this process, green/nature consumption is justified by city expansion, which has activated processes of increasing coastal erosion and loss as well as resource consumption.

The concept of sand/nature urbanisation, investigated in relation to the processes of beautification and coastalisation, has provided a useful lens to understand the complexities of the relationships and interdependencies across the two cases. It has shown how these places must be investigated as complex realities where different interests are in competition, generating different kinds of risks and also activating cascading mechanisms that could be difficult to defuse.

According to the authors, these cases underscore the need for a holistic approach to managing the Italian coast, balancing the needs of tourist development with environmental conservation and the prevention of hydrogeological instability, in order to reduce the risk of misalignment between the image of the country we want to sell and the reality of a territory more and more at risk.

A renewed awareness of territorial values and connections places the Marche and Campania regions in a national and international debate seeking truly effective shared policies at the different governance levels that take into account how humans have already altered certain natural balances. This cannot happen without adopting a multi-scalar perspective. Indeed, the phenomena we have discussed in this chapter move between the local and the global and require an increasing awareness that transcends the national scale.

Notes

1 https://www.mettilatestaaposto.it/spiaggia-di-velluto-parfum
2 See the Georeferenced database of coastal discharges: https://www.arpacampania.it/web/guest/database-georeferenziato-scarichi-costieri

References

Alberico, I., & Petrosino, P. (2014). Territorial evolution and volcanic-hazard, Ischia Island (Southern Italy). *Journal of Maps*, *10*(2), 238–248.

Banca d'Italia (2023). Indagine sul turismo internazionale. https://www.bancaditalia.it/media/notizia/indagine-sul-turismo-internazionale-2023/?com.dotmarketing.htmlpage.language=102

Buchner Niola, D. (1965). *L'isola d'Ischia, Memorie di geografia economica e antropica.* Istituto di Geografia.

Budetta, P., Ducci, D., & Nappi, M. (2013). *Ischia: fattori naturali e paesaggio a rischio.* In E. Petroncelli (Ed.). *Il paesaggio tra rischio e riqualificazione* (pp. 65–76). Liguori.

Casareale, C., Gioia, E., Colocci, A., Marchetti, N., & Marincioni, F. (2023). Perception of the self-exposure to geohazards in the Italian coastal population of the Adriatic basin. In S. D'Amico, & F. De Pascale (Eds.), *Geohazards and disaster risk reduction. Advances in natural and technological hazards research*, vol. 51. Springer. https://doi.org/10.1007/978-3-031-24541-1_3

Cassani, M. (2020). La fiera di Senigallia (1458-1869) tra storiografia e Apoche comunali. *Quaderni del Consiglio Regionale delle Marche*, *307*, 313.

Celata, F., & Gioia, E. (2024). Resist or retreat? Beach erosion and the climate crisis in Italy: Scenarios, impacts and challenges. *Applied Geography*, *169*, 103335. https://doi.org/10.1016/j.apgeog.2024.103335

Celico, P., Stanzione, D., Esposito, L., Formica, F., Piscopo, V., & De Rosa, B. (1999). La complessità idrogeologica di un'area vulcanica attiva: L'isola d'Ischia (Napoli-Campania). *Bollettino della Società Geologica Italiana*, *118*, 485–504.

Cutter, S. L., & Finch, C. (2008). Temporal and spatial changes in social vulnerability to natural hazards. *Proceedings of the National Academy of Sciences*, *105*(7), 2301–2306.

Dawson, K. (2021). Geologising urban political ecology (UPE): The urbanisation of sand in Accra, Ghana. *Antipode*, *53*(4), 995–1017.

De Alteriis, G., & Violante, C. (2009). Catastrophic landslides of Ischia Volcanic Island (Italy). *Geological Society, London, Special Publications*, *322*, 73–104.

Del Prete, S., & Mele, R. (2006). The contribution of historical and archival research in the assessment of landslide susceptibility of Ischia Island. *Rendiconti della Società Geologica Italiana*, *2*, 29–47.

Delizia, I. (1987). *Ischia, l'identità negata*. ESI.

Di Martire, D., De Rosa, M., Pesce, V., Santangelo, M. A., & Calcaterra, D. (2012). Landslide hazard and land management in high-density urban areas of campania region, Italy. *Natural Hazards and Earth System Science*, *12*, 905–926.

Donadio, C., Delogu, M. R., D'Aniello, M., Giovanzanti, A., Guadagno, E., Lämmle, L., Perez Filho, A., Perriello Zampelli, S., Sannino, C., Valente, R., & Donadio, C. (2023). Environmental and geomorphic aspects of urban beaches of Naples, southern Italy. *Bulletin of Regional Natural History (BORNH)/Bollettino della Società dei Naturalisti in Napoli*, *3*(3), 24–45.

Esposito, G., & Matano, F. (2023). A geodatabase of historical landslide events occurring in the highly urbanized volcanic area of Campi Flegrei, Italy. *Earth System Science Data*, *15*, 1133–1149, https://doi.org/10.5194/essd-15-1133-2023

Fitzgerald, A., & Fitzgerald, S. (2008). *Ischia e Capri*. Imagaenaria.

Gambi, M. C. (2014). L'isola d'Ischia: un osservatorio speciale per lo studio del cambiamento climatico globale a mare. In U. Leone, & O. Greco (Eds.), *Ischia Patrimonio dell'Umanità. Natura e cultura* (pp. 71–97). Naples, Doppiavoce.

Gioia, E. (2023). Il rischio alluvione nelle Marche: lezioni apprese dall'evento del 15-16 settembre 2022. "Marca/Marche". *Rivista di storia regionale*, *20*(2023), 70–78.

Gioia, E., & Guadagno, E. (2024). Perception of climate change impacts, urbanization, and coastal planning in the Gaeta Gulf (central Tyrrhenian Sea): A multidimensional approach[J]. *AIMS Geosciences*, *10*(1), 80–106. https://doi.org/10.3934/geosci.2024006

Goffi, G. (2010). *Management delle destinazioni turistiche: sfide per territori e imprese*. Franco Angeli.

Guadagno, E., & Grasso, M. (2022). Le coste in Italia: una questione «frastagliata». *Geotema*, *XXXVI*(69), 24–38

Gugg, G. (2018). Le trame di Casasmicciola (Ischia). Il lavoro culturale (21 August, 2018) https://www.lavoroculturale.org/le-trame-del-terremoto-di-casamicciola-ischia/giovanni-gugg/2018/

ISPRA. (2022). *EcoAtlante*. Retrieved by October 23, 2024, from ecoatlante.isprambiente.it

ISPRA. (2023). *Progetto 'Stato e variazioni delle coste italiane 2020'*. https://indicatoriambientali.isprambiente.it/it/pon/linea-1-linea-3/dinamica-litoranea-erosione-e-avanzamento

ISPRA. (2024). *Cambiamento climatici*. https://www.isprambiente.gov.it/it/attivita/cambiamenti-climatici

ISTAT. (2022). *Classificazione dei comuni in base alla densità turistica*. https://www.istat.it/classificazione/classificazione-dei-comuni-in-base-alla-densita-turistica/

ISTAT. (n.d.). *Turismo*. https://esploradati.istat.it/databrowser/#/it/dw/categories/IT1,Z0700SER,1.0/SER_TOURISM

Lagarias, A., & Stratigea, A. (2023). Coastalisation patterns in the Mediterranean: A spatiotemporal analysis of coastal urban sprawl in tourism destination areas. *Geojournal*, 88, 2529–2552.

Legambiente. (2022). Rapporto spiagge 2022. La situazione ed i cambiamenti in corso nelle aree costiere italiane. https://www.legambiente.it/wp-content/uploads/2022/07/Rapporto-Spiagge-2022.pdf

Marche Tourism Observatory. (n.d.-a). *Total arrivals and stays by type of accommodation, municipality, and province of Marche – Year 2022, Marche Region, Management Control and Statistical Systems Sector on Istat data.* https://statistica.regione.marche.it/Marche-in-Numeri/Turismo

Marche Tourism Observatory. (n.d.-b). *Interactive report – Tourism, Marche region, management control and statistical systems sector on Istat data.* https://statistica.regione.marche.it/Marche-in-Numeri/Turismo

Mazzetti, E. (2006). *Mare. La natura, gli uomini, i porti, le flotte, nella storia, nell'economia, nella cultura.* Guida.

Mele, R., & Del Prete, S. (1998). Fenomeni di instabilità dei versanti in Tufo Verde del Monte Epomeo (isola d'Ischia – Campania). *Bollettino della Società Geografica italiana, 117*(1), 93–112.

Monti Guarnieri, G. (1961). *Annali di Senigallia.* Società Amici Arte e Cultura.

Monti, L., Giulivo, I., Marianelli, P., Mazzarella, A., Esposito, A. C., Di Iorio, P. P., Donadio, C., Putignano, M. L., Toccaceli, R. M., & Gambi, M. C. (2008). *Guida Geologico-Ambientale dell'isola D'Ischia.* Litografia Artistica Cartografica.

Monti, S. (2006). *Geografia e termalismo.* Loffredo.

Munafò, M. (Ed.). (2023). Consumo di suolo, dinamiche territoriali e servizi ecosistemici. Edizione 2023. Report SNPA 37/23. https://www.snpambiente.it/wp-content/uploads/2023/10/Rapporto_consumo_di_suolo_2023.pdf

Pelling, M. (2001). Natural disasters? In N. Castree, & B. Braun (Eds.), *Social nature: Theory, practice and politics* (pp. 170–188). Basil Blackwell Ltd.

Pupazzoni, G. (1975). *Senigallia dopo l'Unità: dalla "Fiera" al turismo.* Tipografia marchigiana.

Ruocco, D. (1988). Ischia: una perla del turismo d'Italia. Studi e ricerche di Geografia, XXI (1), 10–55.

Russo Krauss, D. (2023). L'evoluzione del turismo a Capri: storia di una trasfigurazione che minaccia mito e identità. *Semestrale di Studi e Ricerche in Geografia*, XXXV(2), 153–167. https://doi.org/10.13133/2784-9643/18309

Santoni, G. (2017). *Il terremoto di Senigallia del 1930 e il Piano Regolatore e di Ampliamento della città del 1931. Foto, planimetrie, epigrafi, tabelle e documenti inediti.* Biblioteca Comunale Antonelliana.

Santoni, G., & Morici, R. (2021). *Misa amaro. Piogge intense, inondazioni e alluvioni storiche a Senigallia. Parte seconda - Secolo XX.* Biblioteca Comunale Antonelliana.

Sibilio, G., Russo, A., Vallariello, R., Menale, B., De Luca, P., & De Castro, O. (2015). The past, present and future of thermophilous Cyperus polystachyos Rottb. (Cyperaceae) on the island of Ischia (southern Italy). *Plant Biosystems - An International Journal Dealing with all Aspects of Plant Biology, 149*(5), 933–942. https://doi.org/10.1080/11263504.2014.951713

Sisma ed Emergenza a Ischia. (2023). Rapporto annuale sullo stato della Ricostruzione sull'Isola di Ischia, Ischia. https://sismaischia.it/wp-content/uploads/2023/08/Rapporto-ricostruzione-post-sisma-e-post-frana_31-agosto-2023-con-allegato.pdf

9 The expansion of agrivoltaics in Italy

A sustainable panacea or a new form of green grabbing?

Francesca Peroni, Annachiara Rosati, Virginia Vona, and Daniele Codato

Agrivoltaic technology: a new panacea for climate mitigation?

Among the strategies and interventions to mitigate the effects of climate change, the adoption of renewable energy has emerged as a highly valuable alternative to the extraction and burning of fossil fuels, which significantly contribute to greenhouse gas emissions (IPCC, 2011). Therefore, governments around the world have recognised the potential risks associated with reliance on traditional energy sources and have strategically turned to renewable sources of energy, implementing energy-related policies such as the US Inflation Reduction Act and the European Commission's REPowerEU plan (European Commission, 2022; REN21, 2023).

However, several concerns about renewable energies have emerged, including their impacts on the local environment and communities, land use and landscapes. For example, large-scale land-based solar plants that compete with food production for land allocation (Müller & Pampus, 2023) can alter the aesthetic and visual quality and the sense of place of a landscape (Ferrario, 2018) and cause soil sealing, which, even if it is usually considered reversible in the land take accounting, could alter soil properties (Moscatelli et al., 2022). A key concern investigated by scholars pertains to the commodification of agricultural landscapes in relation to the growth and expansion of renewable energy systems, such as biomass, solar and wind energy. Scheidel and Sorman (2012) recognised a global land rush driven by the need to move away from fossil fuels, fostering land grabs for alternative energies, which demand extensive areas. This issue could be framed in the green grabbing context, which is the global process of acquiring and exploiting natural resources for environmental purposes (Fairhead et al., 2012; Lipari, 2020). A significant case study was reported by Müller and Pampus (2023), who described an example of green grabbing in Brandenburg, Germany. Brokers and investors are leading a rush to grab the largest available areas to install large-scale solar projects. However, German policies are unable to regulate this field, leaving it up to the market to define the rules (Müller & Pampus, 2023). Therefore, while climate achievements appear to form the basis of these practices, the social impacts on local communities living in the area of influence of the project are usually less considered by developers and policy-makers (Müller & Pampus, 2023).

DOI: 10.4324/9781003465164-12

In this framework, among the renewable energy measures, a new system is spreading. Agrivoltaic systems combine food and energy production and are considered a promising solution to contrast competition for land. In such systems, photovoltaic panels are positioned sufficiently above the ground to allow traditional cropping to occur below (Agostini et al., 2021).

Various scientific studies have been conducted to identify the potential of this technology and overcome possible limitations. As reported by Toledo and Scognamiglio (2021), the main issues of agrivoltaic systems are related to agricultural and energy dimensions, that is, reducing shadows on crops and increasing electric energy generation. The authors also underline social issues, such as the acceptance of this technology by citizens, and the landscape dimension; however, little has been reported about these dimensions.

The agrivoltaic issue is well aligned with the concept of social nature in Italy, given the complex and often contradictory ways in which land, resources and technology are used in contemporary societies. This complexity highlights the broader challenge of balancing contemporary energy needs while preserving ecological integrity and promoting environmental justice. From the global to local levels, the "Bel Paese" narrative of beauty, culture and heritage is being repurposed in response to global sustainability agendas and economic forces. This chapter presents some initial insights into the implementation of industrial-scale agrivoltaic energy in Italy, framing it within the debate on green grabbing. In particular, we aim to focus on: (1) the analysis of the scientific literature on agrivoltaics at international scale; (2) the analysis of European and Italian legislation; (3) the analysis of the business developers who have submitted a project to the national Environmental Impact Assessment (EIA), which is mandatory for large agrivoltaic projects.

The chapter is organised as follows: The next section briefly introduces green and landscape grabbing. This is followed by an overview and analysis of the current scientific literature on agrivoltaic systems. The main policies at the European and Italian levels are then discussed, followed by an analysis of the EIA. Finally, we discuss our findings and present our conclusions.

Green and landscape grabbing: framing the issue

Green grabbing derives from land grabbing. According to Borras and Franco (2010), land grabbing refers to large-scale and long-term farmland acquisition through land purchase or leasing, usually for the purpose of agricultural or industrial commodity production. By contrast, green grabbing refers to cases in which the core drivers of these investments are justified by environmental and sustainable agendas (Fairhead et al., 2012), including biodiversity protection, carbon sequestration, the implementation of ecosystem services, ecotourism and, more recently, climate change mitigation and adaptation strategies.

Although the analyses on agrivoltaic system are still in the beginning stages, it could be framed within the debate on green grabbing as demonstrated for other categories of green energy production, such as wind and solar power generation (Siamanta, 2017; Traldi, 2021).

Taking Harvey's (2005) study as a point of departure, the concept of accumulation by dispossession includes the corporatisation and privatisation of public and common goods, such as water. According to Harvey, such mechanisms make profitable use of natural resources (Traldi, 2021).

In this framework, the sun is also considered a common good; hence, when it is integrated into private properties and used for electricity production, it takes the form of a commodity. Therefore, the expropriation or leasing of proprieties for agrivoltaics can also be included among the new accumulation mechanisms described by Harvey: as water, this technology represents a commodification of nature in the form of sun and occupied soil. Finally, it is important to highlight that this process is carried out with the justification that it produces renewable energy to replace fossil fuels and is thus a key strategy to combat climate change.

In addition, the concept of landscape grabbing, which can also be related to large agrivoltaic projects, was recently described by Ciervo and Cerreti (2020). Compared to land grabbing, the focus here is the quick and pervasive landscape transformation in urban and rural parts of developed countries, together with an immaterial change of the sense of place and belonging of local communities. This is often carried out without the acquisition of the lands and due to exogenous and globalisation logics or regenerative sustainability projects that lead to gentrification, which fosters conflicts between exogenous and endogenous promoters and inhabitants.

Scientific literature analysis: an emerging issue

This section presents the state of the art of agrivoltaics phenomenon, via a scientific literature review of studies produced in the last 13 years (last update: February 2024) using the peer-reviewed literature database Scopus. An in-depth review of social science papers was then conducted to determine whether and how topics such as the commodification of nature, green grabbing and landscape values have been addressed. A preliminary analysis showed that, before 2011, articles related to the phenomenon were completely absent. The systematic review considered four keywords: "agrivoltaics", "agrophotovoltaic", "agrovoltaic" and "agriphotovoltaic". The queries were performed in the "Article Title", "Abstract" and "Keywords" fields of the Scopus database, and only papers in English were retrieved. Moreover, it identified 261 peer-reviewed articles published in international scientific journals.

Three metrics were analysed: (i) articles by year, (ii) articles by country and (iii) articles by subject area (Figure 9.1).

According to the first metric, scientific interest in agrivoltaics mainly emerged in 2020, when nine articles were published. This was followed by exponential growth: 45 articles were published in 2021, 63 in 2022 and 97 in 2023. In 2024, 29 articles have already been accepted for publication.

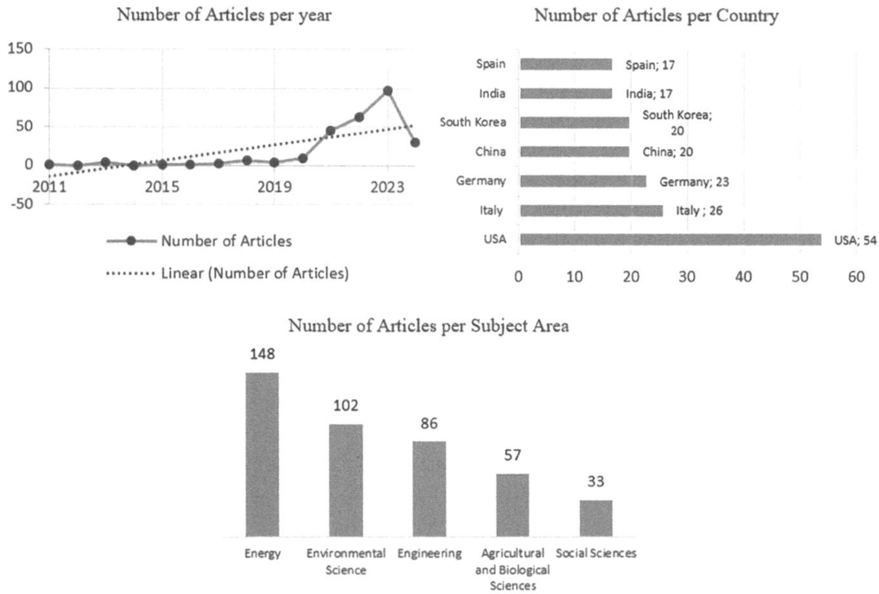

Figure 9.1 Chart illustration of the number of articles by year, country and subject area.

According to the second metric, the United States produced the most articles on the phenomenon (54 articles), followed by Italy (26), Germany (23), China (20), South Korea (20), India (17) and Spain (17).

Finally, agrivoltaic systems were mainly investigated in the subject area of energy (148 articles), followed by environmental science (102), engineering (86), agricultural and biological sciences (57) and social sciences (33). Hence, while agrivoltaics is promoted as an ideal solution for combining energy with food production, the results show that issues related to energy have received much more attention.

For the in-depth analysis of social science studies, we analysed and categorised the articles according to their abstracts, and five macro topics were identified: (i) social acceptance (4 articles), (ii) landscape amenity value (2), (iii) emissions and net zero targets (4), (iv) energy and food production (12) and (v) other (11). Regarding the first and second categories, Hu (2023) addressed important issues of social injustice in large agrivoltaic projects in rural China, which has seen the spread of a phenomenon called "solar extractivism" that leads to macro exploitation of natural resources without considering the well-being of people and nature on-site and is driven by political and economic forces rather than by technological advantages and climate improvements. He also highlights the apolitical character of current investigations on agrivoltaics, focused on technological and economic aspects, while sociopolitical and ecological dimensions are rarely investigated. Hilker et al. (2024) emphasised how local communities in Germany prefer to prioritise photovoltaics on rooftops rather than on the ground, stressing the need for panel installation to respect nature conservation goals. According to Kim et al. (2021), the perception of inhabitants living close to the area of photovoltaic panels over agricultural fields has not always been positive, as they decrease the value of landscapes. Finally, the third macro topic deals with the reduction of carbon emissions through agrivoltaic practices (considered green energy production) to reach net zero goals (Gao et al., 2022; Jamil et al., 2023).

In conclusion, unlike issues related to energy production, the social and cultural aspects of this technology have not been sufficiently investigated.

Agrivoltaic policies: from the European Union to the Italian context

This section examines European and Italian legislation to determine whether it addresses the social implications of agrivoltaic practices, particularly concerning the commodification of nature and green grabbing. To date (February 2024), there is no single accepted definition of "agrivoltaics", and there is a lack of European legislation on this technology. In Spain and Poland, for example, agrivoltaics are still considered traditional ground-mounted photovoltaics, and there are no distinctions between their project authorisation procedures, which differ in their restrictions by country (Chatzipanagi et al., 2023). In the Netherlands, solar energy on roofs is considered more appropriate than installation on agricultural lands, which should be considered a last option (Netherlands Enterprise Agency, 2021). Germany, France and Italy are the three EU countries with the most regulations on this new technology. Germany recognised agrivoltaic systems in 2021 and relevant guidelines have been developed significantly (Fraunhofer Institute for Solar Energy Systems ISE, 2022). In France, the first guidelines allowed the co-location of agricultural and photovoltaic production since 2021 (Article L123-1 – Code de l'urbanisme, 2021), while in 2023, the different categories were defined, including greenhouses equipped with solar panels.

The fragmentation of Italian legislation poses serious challenges (Table 9.1). According to Decree-Law No. 77/2021 (Decreto Legge 31 Maggio, 2021), an early legislative description of agrivoltaics was offered as a distinct category from traditional photovoltaic systems. In addition, the National Recovery and Resilience Plan (PNRR) allocated 1.1 billion euros for agrivoltaic systems and specified the requirements for incentive eligibility (Italia Domani & Piano Nazionale di Ripresa e Resilienza, 2021). Decree-Law No. 17/2022 defines the Simplified Administrative Procedure[1] and access to subsidies for agrivoltaic systems (Decreto Legge 1 Marzo, 2022).

In June 2022, the Guidelines on Agrivoltaic installations defined the minimal qualities and standards to classify an agrivoltaic system to be eligible for PNRR subsidies (Ministero della Transizione Ecologica – Dipartimento per l'energia, 2022). The main requirements are as follows: (A) the arrangement of space and technology decisions that ensure the mutually beneficial synergy of electricity and crops; (B) the maintenance of the uninterrupted flow of pastoral and

Table 9.1 Summary table of Italian laws that refer to agrivoltaics

Name of the law	*Law/ Guidelines*	*Description*
Decree Law n. 77/2021	L	First legal definition of agrivoltaics
Legislative Decree n. 199 (November 8, 2021)	L	Regulations for receiving the subsidies provided by the National Recovery and Resilience Plan
National Recovery and Resilience Plan	L	1.1 billion € are allocated for agrivoltaic system installations
Decree Law n. 17/2022 (Clauses 9 and 11)	L	The Simplified Administrative Procedure and access to subsidies for agrivoltaic plants built in suitable areas
***Guidelines on Agrivoltaic Plants* | Ministry of the Environment and Energy Security (June 2022)**	G	Features and minimum requirements for agrivoltaics to obtain eligibility for the National Recovery and Resilience Plan subsidies
Decree Law n. 13/2023	L	More legislative simplifications for agrivoltaics Updates: Converted, with amendments, by Law No. 41 (April 21, 2023) | (text not consistent with the Guidelines of Ministry of the Environment and Energy Security)
***Agrivoltaic systems: integration of agricultural activities and photovoltaic systems*/Italian Standards Body Reference Practice (March 9-April 12, 2023)**	G	Addition of some parameters to complete the Guidelines of the Ministry of the Environment and Energy Security
Consultation on working rules for agri-voltaic systems* | Manager of Energy Services *(5–21 July 2023)	G	Definition of future standards designed for the National Recovery and Resilience Plan-fundable agrivoltaics, based on the Ministry of the Environment and Energy Security Guidelines
"Suitable areas" decree (draft pending review) **July 2023**	L	Definition of suitable and unsuitable areas for agrivoltaics
Decree "innovative agrivoltaic" (pending validation) **December 21, 2023**	L	Further indications for the use of the National Recovery and Resilience Plan funding to promote innovative agrivoltaics

agricultural activities; (C) the introduction of innovative integrated solutions to maximise the energy and agricultural performance; (D) the existence of a monitoring system to confirm the effects on crops, water savings, agricultural productivity for the several crop varieties, and the ongoing operations of the involved farms; and (E) the existence of a monitoring system that satisfies criteria D and allows verification of the restoration of soil fertility, microclimate and climate change resilience. Furthermore, a photovoltaic system in an agricultural area is defined as "agrivoltaics", when it meets conditions A, B and D.2 (i.e., continuation of agricultural activities), and it is defined as an "advanced agrivoltaic system" when conditions A, B, C and D are met.

Decree-Law No. 13/2023 (Decreto Legge 24 Febbraio, 2023) defines other regulatory simplifications; however, this law is not aligned with the previous definitions, such as those in previous guidelines. These latter definitions were picked up and re-proposed in the Italian Standards Body Reference Practice in 2022 along with a few new parameters (Una Prassi Di Riferimento per i Sistemi Agrivoltaici, 2023). Specifically, an agrivoltaic system's agricultural output should be at least 70%; the yield reduction should not exceed 30% when compared to standard production conditions – that is, an open field without agrivoltaics; and the inclusion of a control area extending between 1,000 and 2,000 m^2 is recommended.

To gather valuable data for the development of future standards, the Manager of Energy Services started a new discussion on the topic in July 2023 by sending out a consultation document regarding the rules of operation for agrivoltaic systems that should be eligible for PNRR resources (GSE, 2023).

In December 2023, a new decree was signed that encouraged the spread of "innovative" agrivoltaics. This measure, which has already been approved by the European Commission, supports the goal of installing at least 1.04 GW of advanced agrivoltaic systems by 30 June 2026 using PNRR funds through two measures: (i) the granting of capital contributions of up to 40% of eligible costs and (ii) an incentive tariff on net electricity production fed into the grid. To date, the law has yet to take effect.

One of the main issues in Italy related to agrivoltaics is where to implement these systems. *De facto*, no law consistently specifies suitable and unsuitable areas for deployment. The current regulation defining the standards is included in the Ministerial Decree of September 2010 and in Decree-Law No. 13/2023. However, this last Decree-Law does not distinguish between photovoltaic and agrivoltaic systems. As reported by Pileri (2022), the locations identified by the law – without following any scientific basis – are belt areas around or along previously installed systems and other types of infrastructure. As a consequence, it seems that the legislator recognises these places' "poor" level of ecosystem sensitivity or anthropisation, and it allows for pre-emptive exploitation.

It is important to note that the previous Ministerial Decree established specific objective and technical standards related to preserving cultural, landscape and environmental heritage. These standards were applied to identify areas deemed unsuitable for certain purposes, such as agricultural areas touched by high-quality agricultural and food production. According to the draft decree dated July 2023, these areas are suitable only for the installation of advanced agrivoltaic systems. Moreover, the Italian legislation does not address what kinds of crops are allowed in agrivoltaic systems. This matter should be regulated to prevent the extinction of local productions recognised and protected by protected geographical indications or quality assurance labels that might not work well with agrivoltaics (Pileri, 2022). A list of possible agrivoltaic crops has only been provided in the Ministry of Environment and Energy Security's guidelines, which were based on some scientific studies.

An additional issue emerged from interactions between the actors involved, such as energy companies and farmers, which could result in green grabbing. As already observed in the past, financial speculation arising from the rush to photovoltaics, as well as a lack of clear legislation, has hidden the phenomenon of the eco-mafia, that is, eco-crimes and negative impacts on the territory and environment of activities perpetuated by organised crime. This scarce attention to eco-mafia infiltration, social conflict and the voices of local communities has also been observed in agrivoltaic projects (Pileri, 2023).

Who is behind large agrivoltaic projects: the national environmental impact assessment analysis

As discussed in previous sections, both the normative sphere and scientific research concerning agrivoltaic projects have mainly focused on climate, energy and agricultural issues, while social dimensions, such as the transformation of landscapes, power relationships among stakeholders and involvement of local communities, have been marginalised. Therefore, in this section, we investigate the characteristics of the project developers, that is, the proponents of large agrivoltaic projects. We refer to projects with generating capacities of more than 10 MW for which authorisation on agricultural land is subjected to the submission of a national EIA and are evaluated by the Technical Commission for the EIA of PNRR-PNIEC projects (Piano Nazionale Integrato per l'Energia e il Clima [Integrated National Plan for Energy and Climate]). Even if the EIA focuses on assessing these types of projects, as well as their environmental impacts and compensation, we aim to investigate the project developer to give an overview of its connection and relationships with the local territory under the influence of the planned project. In fact, according to legislation, two types of actors can access PNRR funds for agrivoltaic projects: the agricultural enterprise (single or associated) or a temporary association of companies (TAC), which usually includes a company from the energy sector and where at least one must be from the agricultural sector (Pennazio & Ferrero, 2023).

First, we downloaded (last update: January 2024) the lists of all EIAs of the PNIEC-PNRR (in progress or completed) from the Ministry of Environment and Energy Security website (Ministero dell'Ambiente e della Sicurezza

Energetica, 2024). These lists contained the following information: (i) a description of the project, including the type, the power generating capacity and administrative area; (ii) the business name of the project developer; and (iii) the starting date of the procedure of EIA. Then, we cleaned the dataset and selected all projects containing the words "agrivoltaics", "agro-voltaic", "agri-photovoltaic" or similar declinations in the description. Overall, we found 744 EIA procedures. The business names of the developers were inserted into the Aida (Analisi Informatizzata delle Aziende Italiane [Computerised Analysis of Italian Companies]) database, which provides personal and production information and balance sheets for all active Italian joint-stock companies (Bureau van Dijk, 2014). We downloaded all available information on EIA developers and reviewed the resulting list to exclude possible errors due to ambiguity (i.e., companies sharing the same name). Out of 662 developer companies, we selected 593, which is about 90%, for a total of 661 EIAs.

Overall, 356 (61%) out of 584 companies with information in the classification of economic activities column, that is, the column titled as Ateco 2007, deal with production, commercialisation, building and consulting regarding energy, many others are generic consulting and project services or holding companies, and only five deal directly with agriculture. From a geographic perspective, Figure 9.2 shows the number of projects under EIA per province and the provinces of the legal office addresses of the developers. The figure highlights a territorial mismatch between the major distribution of large agrivoltaic projects in the southern provinces (mainly Puglia, Sicily and Sardegna) and the localisation of developers concentrated mainly in Milan (201), Rome (146) and Bolzano (73). It is worth noting that operational addresses, where the companies carry out their activities, are indicated by only 103 companies. It is also interesting to examine the origin of the shareholders, as for 219 companies (out of 585 with this information), all shareholders are from foreign countries, mainly Germany, Spain, Luxembourg and the Netherlands.

Lastly, by filtering the EIAs according to their starting date, that is for years 2021 (102 EIAs procedures), 2022 (278) and 2023–24 (281), we analysed the year of establishment of the companies, as well as the number of employees and the sales revenues of the last available year. The results show that almost all developers can be divided into two large groups. The larger group is composed of companies that appear to have been "built ad hoc" for the projects under EIA, probably as special purpose vehicle companies: 547 out of 593 were established since 2019, only two years prior to the establishment of the PNIEC; among these, only five report the number of employees and 26 report sales revenues for the last available year. The other group contains 31 "more consolidated" companies that report the number of employees and/or sales revenues. The results of these preliminary analyses highlight the need to pay greater attention to the companies proposing the projects from the energy sector, as well as how they will establish TACs and agreements with farmers.

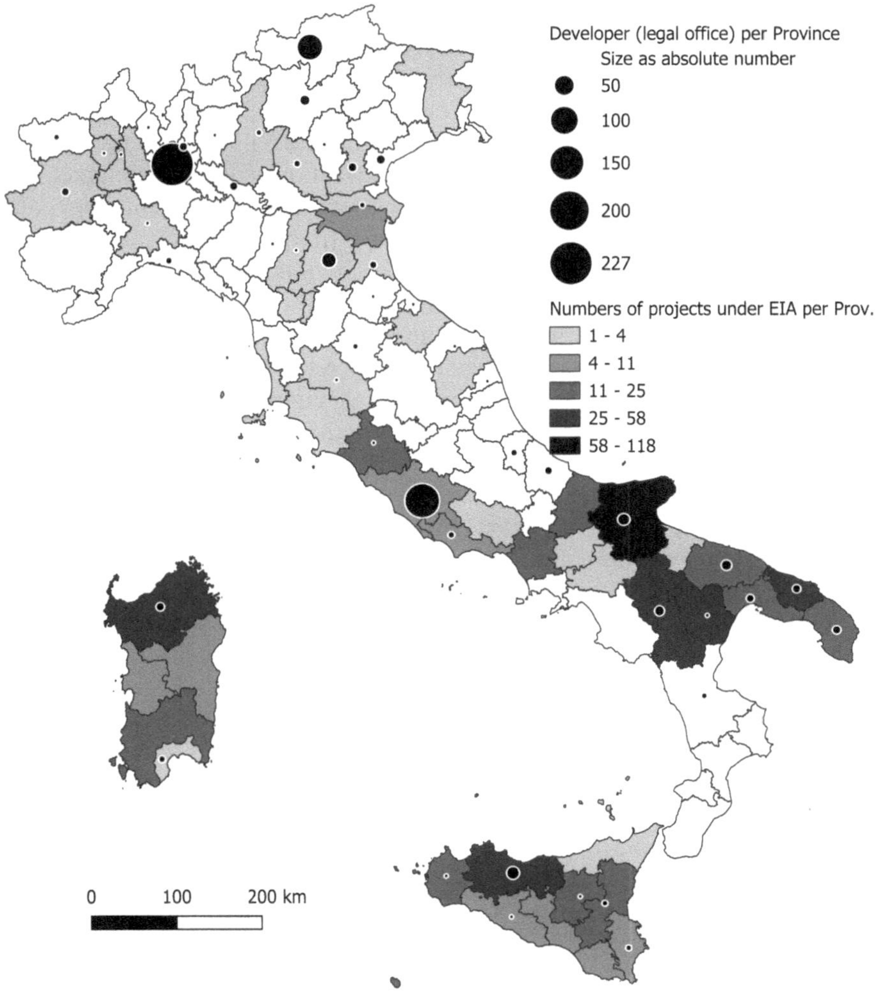

Figure 9.2 Distribution of the number of project developers (legal address) and EIA projects per Province.

Discussion

According to our scientific literature review, from a climate, agronomic and energetic perspective, agrivoltaic technology appears to be considered a "sustainable panacea" that could solve the trade-offs between decarbonisation and energy production, competition for land and agricultural production. However, the consequences of the spread of this new technology, particularly in industrial-scale projects, have not yet been well investigated, such as its social and environmental impacts on local inhabitants and landscapes, and there is a risk of perpetuating some of the negative dynamics investigated in other forms of renewable energy production (Müller & Pampus, 2023).

Moreover, we suggest that large-scale agrivoltaic systems could be considered a new form of commodification of nature, in which the older processes of appropriation are reproduced and innovatively transformed, spreading the phenomenon of solar extractivism (Hu, 2023; Traldi, 2021). These new dynamics involve new (re)configurations of actors and relationships (mainly large energy companies and some agricultural companies), the reinvention of legal and market processes (new state subsidies to support the development of this technology) and the development of novel justificatory discourses (mitigation of climate change).

From this perspective, the EIA analysis highlights the beginning of a process of accumulation of capital (i.e., land) by energy companies that look to be far away from farmers and local communities and which relationships should be monitored to avoid disproportionate power relationships. In addition, the new value of agricultural lands obtained through agrivoltaic projects is creating new financial opportunities for investors (national and international), local elites and the state. A balance of power among companies, farmers and local communities should be ensured, as well as the fulfillment of the guarantees to minimise or avoid negative impacts presented in the EIAs, both during the operational and decommissioning phases.

As reported in our policy analysis, although Italian energy national policies currently support this system, several new regulations that appear to promote this new field are being implemented; however, several territorial concerns, such as social and landscape impacts, remain unexplored and require further investigation and attention from policy-makers. Moreover, as with the previous rush to photovoltaics, an eco-mafia phenomenon could also occur in agrivoltaic projects (Pileri, 2023). However, policies are in continuous development, and the European Joint Research Centre has recommended including issues such as property security, attention for rural communities and preventing greenwashing (Chatzipanagi et al., 2023).

It is recognised worldwide that the transition from fossil fuels is a priority and that renewable energy projects represent one of the main paths. However, stakeholders' different visions of nature (e.g., nature as clean energy or as rural landscapes) and their different interests (e.g., fighting climate change vs. green extractivism) also form the basis of recent conflicts between environmental associations that are expected to have similar visions of how to balance energy transition and landscapes conservation (Malcevschi, 2023). These conflicts highlight that transition is not an unquestionable and apolitical matter but that there are various visions of transition. The dispute is more alive than ever and should be considered in all arenas, from academia to politics and stakeholders. Moreover, future studies should investigate the relationship between the challenges and crises affecting European agriculture and the spread of the agrivoltaic phenomenon. Indeed, from this perspective, many farmers and agricultural owners could be attracted by the new investments proposed by the agrivoltaic industry. As a consequence, it will also be important to analyse whether the installation of this new technology involves the complete alienation of the land by farmers or whether it aims to restructure the rules of access to resource use and management.

Conclusion

This chapter aimed to provide initial insights into the implementation of large-scale agrivoltaic energy systems in Italy, framing them within the discourse on green and landscape grabbing. The results of the three analyses highlight how this new technology is considered a "sustainable panacea", mainly by the energy sector, justified by the production of renewable energy to replace fossil fuels. Thus, this technology is recognised as an important strategy for combating climate change. However, the processes beyond the spread of this system, such as the relationships between the different actors involved (mainly private energy companies), the leasing or purchase of agricultural land, the modification of landscapes and the impacts on local communities – which have not been sufficiently investigated – could be attributed to the green grabbing dynamics and a process of commodification of natural resources. Moreover, the lack of clear legislation that regulates this field is likely to exacerbate the green grabbing phenomenon, leaving it to the market to define the rules.

In conclusion, the expansion of agrivoltaics in Italy offers new opportunities for further exploration within human geography through the theoretical framework of social nature. Future research should explore the role of agrivoltaics not only as a solution to the climate crisis but also as a potential driver of new socio-environmental conflicts. This reinforces the importance of integrating social, ecological and technological perspectives to ensure just and sustainable outcomes. By addressing these challenges, we can move beyond the *Bel Paese* narrative towards more inclusive and equitable forms of environmental management.

Note

1 The Simplified Administrative Procedure is a permission designed to encourage the spread of renewable energy generation systems. The declaration, which is made up of several reports and project documents, must be given to the municipality at least thirty days prior to the start of work.

References

Agostini, A., Colauzzi, M., & Amaducci, S. (2021). Innovative agrivoltaic systems to produce sustainable energy: An economic and environmental assessment. *Applied Energy*, *281*, 116102. https://doi.org/10.1016/J.APENERGY.2020.116102

Article L123-1 – Code de l'urbanisme. (2021). https://www.legifrance.gouv.fr/codes/article_lc/LEGIARTI000043977726

Borras, S. M., & Franco, J. C. (2010). From threat to opportunity? Problems with codes of conduct for land grabbing. *Rethinking Food Systems: Structural Challenges, New Strategies and the Law*, 147–162. https://doi.org/10.1007/978-94-007-7778-1_7/COVER

Bureau, van Dijk. (2014). *AIDA Guida*.

Chatzipanagi, A., Taylor, N., & Jaeger-Waldau, A. (2023). *Overview of the potential and challenges for Agri-Photovoltaics in the European Union*. EUR 31482 EN, Publications Office of the European Unione, Luzemburg. https://doi.org/10.2760/208702

Ciervo, M., & Cerreti, C. (2020). Landscape grabbing. A new concept for geographical analysis? *Geotema*, Supplemento 2020 (July), 15–18.

Decreto Legge 1 Marzo 2022, n. 17. (2022). https://www.gazzettaufficiale.it/eli/id/2022/03/01/22G00026/sg

Decreto Legge 24 Febbraio 2023, n. 13. (2023). https://www.gazzettaufficiale.it/eli/id/2023/02/24/23G00022/sg

Decreto Legge 31 Maggio 2021, n. 77. (2021). https://www.gazzettaufficiale.it/eli/id/2021/05/31/21G00087/sg

European Commission. (2022). *REPowerEU: Affordable, secure and sustainable energy for Europe*. https://commission.europa.eu/strategy-and-policy/priorities-2019-2024/european-green-deal/repowereu-affordable-secure-and-sustainable-energy-europe_en#next-steps

Fairhead, J., Leach, M., & Scoones, I. (2012). Green Grabbing: A new appropriation of nature? *Journal of Peasant Studies*, *39*(2), 237–261. https://doi.org/10.1080/03066150.2012.671770

Ferrario, V. (2018). I nuovi paesaggi delle energie rinnovabili. Occasioni di conflitto e spazi per l'integrazione. In B. Castiglioni, C. P. Santacroce, C. Quaglia, & A. Dal Pozzo (Eds.), *Il paesaggio tra conflittualità e integrazione. Materiali da un'esperienza formativa* (pp. 33–41). Cleup.

Fraunhofer Institute for Solar Energy Systems ISE. (2022). *Agrivoltaics: Opportunities for agri-culture and the energy transition*. www.ise.fraunhofer.de

Gao, Y., Zhang, Y., Zhou, Q., Han, L., Zhou, J., Zhang, Y., Li, B., Mu, W., & Gao, C. (2022). Potential of ecosystem carbon sinks to "neutralize" carbon emissions: A case study of Qinghai in west China and a tale of two stages. *Global Transitions*, *4*, 1–10. https://doi.org/10.1016/J.GLT.2022.08.001

Gestore Servizi Elettrici (GSE). (2023). *Sistemi agrivoltaici. Consultazione sulle regole operative*. https://www.gse.it/sostenibilita_site/Stakeholder_site/consultazioni_site/Documents/SISTEMI%20AGRIVOLTAICI%20-%20Documento%20di%20consultazione%20sulle%20regole%20operative.pdf

Harvey, D. (2005). *The new imperialism*. Oxford University Press.

Hilker, J. M., Busse, M., Müller, K., & Zscheischler, J. (2024). Photovoltaics in agricultural landscapes: "Industrial land use" or a "real compromise" between renewable energy and biodiversity? Perspectives of German nature conservation associations. *Energy, Sustainability and Society*, *14*(1), 1–15. https://doi.org/10.1186/S13705-023-00431-2/FIGURES/6

Hu, Z. (2023). Towards solar extractivism? A political ecology understanding of the solar energy and agriculture boom in rural China. *Energy Research & Social Science*, *98*, 102988. https://doi.org/10.1016/J.ERSS.2023.102988

IPCC. (2011). Summary for policy makers. In O. Edenhofer, R. Pichs-Madruga, Y. Sokona, K. Seyboth, P. Matschoss, S. Kadner, T. Zwickel, P. Eickemeier, G. Hansen, S. Schlömer, & C. von Stechow (Eds.), *IPCC special report on renewable energy sources and climate change mitigation*. Cambridge University Press.

Italia Domani, & Piano Nazionale di Ripresa e Resilienza. (2021). *Sviluppo agro-voltaico*. https://www.italiadomani.gov.it/it/Interventi/investimenti/sviluppo-agro-voltaico.html

Jamil, U., Bonnington, A., & Pearce, J. M. (2023). The agrivoltaic potential of Canada. *Sustainability*, *15*(4), 3228. https://doi.org/10.3390/SU15043228

Kim, T. H., Chun, K. S., Yang, S. R., Brugarolas Molla-Bauza, M. M., & Martinez-Carrasco, L. (2021). Analyzing the impact of agrophotovoltaic power plants on the amenity value of agricultural landscape: The case of the Republic of Korea. *Sustainability*, *13*(20), 11325. https://doi.org/10.3390/SU132011325

Lipari, S. (2020). View of industrial-scale wind energy in Italian southern Apennine | Scienze del Territorio. *Scienze Del Territorio*, *8*, 154–169.

Malcevschi, S. (2023). *Paesaggio ed energie rinnovabili in conflitto. Aspetti critici della valutazione*. http://www.catap.eu/wp-content/uploads/2023/01/Riv.VA_Malcevschi_Paesaggio-e-rinnovabili_preliminare230126.pdf

Ministero dell'Ambiente e della Sicurezza Energetica. (2024). *On-going procedures – Environmental assessments and authorizations*. https://va.mite.gov.it/en-GB/Procedure/ProcedureInCorso

Ministero della Transizione Ecologica – Dipartimento per l'energia. (2022). *Linee Guida in materia di Impianti Agrivoltaici*. Ministero della Transizione Ecologica.

Moscatelli, M. C., Marabottini, R., Massaccesi, L., & Marinari, S. (2022). Soil properties changes after seven years of ground mounted photovoltaic panels in Central Italy coastal area. *Geoderma Regional*, *29*(December 2021), e00500. https://doi.org/10.1016/j.geodrs.2022.e00500

Müller, K., & Pampus, M. (2023). The solar rush: Invisible land grabbing in East Germany. *International Journal of Sustainable Energy*, *42*(1), 1264–1277. https://doi.org/10.1080/14786451.2023.2260009

Netherlands Enterprise Agency. (2021). *Netherlands enterprise agency*. https://english.rvo.nl/

Pennazio, R., & Ferrero, E. (2023). I sistemi agrovoltaici tra cambiamento climatico e nuove sfide per l'imprenditore agricolo contemporaneo. *Alimenta*, *2*, 355–389. https://iris.uniupo.it/handle/11579/163843

Pileri, P. (2022). *Questa corsa alla pannellizzazione fotovoltaica non fa il bene dei suoli agricoli*. Altreconomia. https://altreconomia.it/questa-corsa-alla-pannellizzazione-fotovoltaica-non-fa-il-bene-dei-suoli-agricoli/

Pileri, P. (2023). *La corsa ai suoli agricoli dei nuovi "lupi solari": una speculazione da fermare*. Altreconomia. https://altreconomia.it/la-corsa-ai-suoli-agricoli-italiani-dei-nuovi-lupi-solari-una-speculazione-da-fermare/

REN21. (2023). *Renewables 2023 global status report collection, global overview*. https://www.ren21.net/gsr-2023/modules/global_overview/

Scheidel, A., & Sorman, A. H. (2012). Energy transitions and the global land rush: Ultimate drivers and persistent consequences. *Global Environmental Change*, *22*(3), 588–595. https://doi.org/10.1016/J.GLOENVCHA.2011.12.005

Siamanta, Z. C. (2017). Building a green economy of low carbon: The Greek post-crisis experience of photovoltaics and financial "green grabbing". *Journal of Political Ecology*, *24*(1), 258–276. https://doi.org/10.2458/V24I1.20806

Toledo, C., & Scognamiglio, A. (2021). Agrivoltaic systems design and assessment: A critical review, and a descriptive model towards a sustainable landscape vision (three-dimensional agrivoltaic patterns). *Sustainability*, *13*(12), 6871. https://doi.org/10.3390/SU13126871

Traldi, M. (2021). Accumulation by dispossession and green grabbing: Wind farms, lease agreements, land appropriation in the Brazilian semiarid. *Ambiente & Sociedade*, *24*, e00522. https://doi.org/10.1590/1809-4422ASOC20200052R2VU2021L4TD

Una Prassi Di Riferimento per i Sistemi Agrivoltaici. (2023). https://www.uni.com/una-prassi-di-riferimento-per-i-sistemi-agrivoltaici/

Part III

More-than (peri)urban natures

10 Urban socionatures and climate strategies

The cases of three medium-sized cities in Northern Italy

Marco Tononi

Urban natures from sustainability to climate strategies

During the 1980s, a new interest in natural elements emerged with the ideas of sustainable development and urban sustainability arising from the rediscovery of nature and its benefits in overpopulated and contaminated industrial cities. This interest continues with the production of urban natures in experiments to make cities more resilient to climate change. An interesting issue that has appeared in different research fields is the political and sociocultural aspects of natures in and of cities. For this reason, in this chapter, some fundamental elements for the analysis of urban natures start from critical neo-Marxism (Smith, 1984; Harvey, 1993), political ecology (Bryant & Bailey, 1997; Forsyth, 2003; Robbins, 2004) and the nature–society hybridity of human and non-human subjects (Whatmore, 2002). The critical points of view in Castree and Braun (2001) constitute a social approach to nature. From this perspective, as the urban political ecology (Desfor & Keil, 2004; Heynen et al., 2006) argues, the production and reproduction of nature is related to social dynamics. Nature is considered not only a separate external element to be protected or managed, but is also part of socioecological interactions that produce spatial results. The different socionatures (Swyngedouw, 1996) inside urban spaces, generated in different moments by the urban metabolism, represent different interactions between society and nature and its effects.

Industrial cities in the Western world consider nature a resource for the industrial metabolism and produce infrastructure and industrial areas to exploit it. At the same time, this industrial metabolism influences the planning of all parts of cities by producing infrastructure or residential areas based on these socio-economic and ecological relations. The environmental problems caused by pollution, incidents or disasters forced cities, starting in the 1970s, to reconsider their approach to nature and use of resources. During cities' industrial growth, nature was not absent but confined to being a mere resource useful only for capitalistic exploitation.

However, there is nothing unnatural in cities, as Harvey (1993) said about New York. Nature is part of cities, and the emergence of sustainability must be interpreted as a way to produce and reproduce urban natures.

DOI: 10.4324/9781003465164-14

In the 1980s, research on urban sustainability along with the reintroduction of nature and the associated environmental issues have become new key elements of urban development in the politics and practices of cities. Wheeler and Beatley (2004) explained well the role of sustainability in the redesign of cities and the production of urban natures; urban sustainability studies started with some pioneer authors on these themes, such as Reclus, Howard and Mumford, who worked on the planning of urban spaces with new attention to environmental problems. As in northern European cities, this is a tendency that nowadays involves urban development at the global level both in America (Beatley, 2012) and Asia (Caprotti & Yu, 2017). Nevertheless, some critical aspects have emerged. Considering the development of urban sustainability issues to be only a technical approach to urban environmental policies and practices obliterates the sociocultural dynamics and historical interactions between society and nature within the specific territorial contexts. Some urban sustainability practices introduce new methods and techniques of natural resource management without overcoming the division between nature and society.

Braun (2005) focuses on a key element concerning this discussion of the urban environment, considering the theory of planetary urbanisation (Brenner, 2014; Lefebvre, 1970) and the interscalar properties of environmental issues in the Anthropocene (Ernstson & Swyngedouw, 2018). In particular, the reproduction of urban natures within urbanisation processes is based on the interrelation of several spatiotemporal scales and socioecological forces, and overrating the local scale as the only fair level of application might be wrong. Research on climate change considers urban spaces to be ecologically interconnected at a global level, while sustainability studies in many cases focus on the local as the only important scale. Kaika et al. (2023), in *Turning up the Heat: Urban Political Ecology for a Climate Emergency*, propose the concept of extended urbanisation as a "unifying field" that, without a hierarchised relationship between the core and the periphery, starts from the margins, still considering the core, and expands the concept of marginality beyond classical spaces and actors. Nature plays a role in these processes of extended urbanisation, producing new socioecological relationships between the cores and peripheries. Urban policies in climate change times are based on experimental approaches, which means reproducing some experiments in different urban areas to improve climatic performance, with some critical aspects to be considered (Broto & Bulkeley, 2013).

The global tendencies to deal with climate change at the urban level have introduced experiments in urban spaces as best practices for the urban environment, in which nature can be seen as a technical, nature-based applied solution. Some authors (Beatley, 2012; Caprotti & Yu, 2017) clearly show that the solution of reimagining cities or some of their parts as ecocities, and their possible definition as smart, green, resilient, sponge, is today a global issue in an increasingly urbanised world. This new centrality of nature, from a Urban Political Ecology point of view, needs to be investigated deeply, reconsidering, on the one hand, the political consequences and the inequality produced by the socioecological metabolism, and, on the other hand, the hybridity of human and non-human nature inside urban processes (Braun, 2005; Connolly, 2019; Gandy, 2005, 2022a, 2022b; Tzaninis et al., 2021).

Following this theoretical scheme, the analysis proposed here focuses on the new (re)production of nature in the urban climate strategies promoted and realised in three Italian cities: Brescia, Bergamo and Mantova. This study investigates the plans and documents produced by these cities' climate urban projects and the points of view, practices and actions of other stakeholders: environmental associations, experts and active citizens. The production of new urban natures in these projects needs to be explored to understand the different ideas of nature, how these fit into the socioecological context characterised by an industrial past and present, the emergent hybridity between social and ecological elements, the unequal distribution of the consequences and role of democracy and citizen participation in the process of urbanisation of nature and the use of nature in rebranding the image of an industrial city. To do this, a content analysis of the different documents published about urban climate strategies in the case studies has been conducted, focusing on the structure of the strategies, resources, main environmental and climate issues, actions promoted and the peculiarities of each strategy. These points were then discussed in depth in interviews with two environmental associations and two municipal technicians and tested with participation in initiatives conducted by the three cities, including public presentations of the strategies and participatory actions in Brescia.

(Re)production of urban natures: socionatures in climate change

Planetary urbanisation (Brenner, 2014; Lefebvre, 1970) or extended urbanisation (Kaika et al., 2023) is a core element of the new Anthropocene era and one of the most evident dynamics that makes clear the impact of human society on Earth. Since 2010, the urban population has exceeded the rural population; over 40% comprise urban settlements with under 300,000 inhabitants,[1] and urban processes also influence suburban and rural areas as the main drivers of human interaction within the Earth's system. To meet the challenges launched by the sustainable development goals, particularly goal 11, sustainable cities and communities, and to deal with climate change at the urban level, it is necessary to work on medium and small urban settlements and not only on the biggest cities (Häußler & Haupt, 2021). The experimental cities approach (Caprotti & Cowley, 2017; Evans et al., 2017), which emerged in dealing with climate change (Broto & Bulkeley, 2013), promotes an innovative way to intervene in the urban environment, involving stakeholders in experimental projects. Indeed, urban management today is strictly influenced by the capacity and possibility of attracting funds and investments from outside. This experimental approach is very common and has been diffused in a period in which different stakeholders, administrations, associations, research institutes and enterprises participate in grants that also reward smart, green, resilient or climate-friendly actions in small-and medium-sized cities. Nature is one of the most important themes of these calls for grants, which allow cities to obtain extra funding at the European or national level (Negreiros & Falconer, 2021). Today, the European Green New Deal[2] is an extraordinary funding plan based on the promotion of ecological transition. Due to new opportunities for investment and funding, small-and medium-sized Italian cities are also paying particular attention to the environmental transition and reconsidering the role of nature in their plans and policies.

In this new centrality of urban natures, green has become "good" (Angelo, 2020) in the process of environmental transition starting from the past industrial city, where nature was often polluted or pollutants were diffused with negative considerations, and arriving at today's climate strategies, where nature is the basis of new practices considered as a positive intervention in regreening industrial cities. At a certain moment, nature became a good element of urban regeneration. However, the reintroduction of nature based on new urban sustainability projects and climate strategies in urban contexts in many cases does not consider that nature was already present inside the processes of industrial cities in many forms (Angelo, 2020; Gandy, 2022b). This approach produces new, promoted as good, natures with sustainability and resilient solutions, such as rain gardens, green roofs, new urban forests, and other nature-based solutions, and forgets some old, sometimes perceived as negative, natural elements as part of the industrial urban past considered affected by past human activity and at least non-natural, such as canals, mining sites or natures such as gardens or uncultivated and rural areas, spaces left out of urbanisation processes (Gandy, 2022b). These socionatures are territorial evidence of different socioecological relationships produced during the development of cities, and the moral connotation as good or not does not represent their value. The ecological transition that territories are experiencing today must recognise these natures and be based on them.

European cities aligned with European green politics, such as the Covenant of Majors,[3] Green European Capital, Green Leaf,[4] and many other programmes, frequently hide or forget old and other natures and plan new green solutions. With the European Green Deal and its National Plan of Recovery and Resilience (NPRR), Italy also has a Plan of Ecological Transition.[5] It is precisely these plans and opportunities that encourage the adoption of climate strategies, which is why city administrations are now interested in taking advantage of new funding opportunities that address environmental problems and conflicts, although in many cases, they hide them and present new images of cities based on green projects and designs, rebranding the image of industrial cities characterised by their past.

In winter 2024, the Po Valley in Italy, where the three case studies are located, is under the spotlight as one of the areas with the highest levels of air pollution not only in Europe[6] but also in the world; some politicians in these cities protest, accusing these data of not representing the real situation. It is not necessary to enter into the debate about the authority of the analysis.

Probably, some consideration of the effect of the use of ranking could be done. Therefore, it is interesting to see how politicians in the media need to defend the new green transitions proposed by their cities, which are starting to change outside perceptions of their cities. New ecological plans for urban areas must deal with an urban metabolism that has an impact on the socioecological system without a boundary between urban, suburban and rural areas. For these reasons, the Po Valley, a complex urbanised and industrial region, needs a multiscalar approach in which different administrative levels, from local to regional, work together to promote an alternative model of development. On the contrary, today's administrations, within or between levels, adopt, in some cases, conflicting policies that, on the one hand, tend to reproduce economic patterns of environmental exploitation in the name of economic growth and, on the other hand, promote sustainable solutions, resulting in a kind of green split personality.

Natures inside urban climate transitions: Brescia, Bergamo and Mantova

The following sections of the chapter analyse how three cities in the Po Valley deal with climate change issues by adopting an experimental approach in urban environmental policies. The three cities, with 50,000 to 200,000 inhabitants, were chosen as medium-sized municipalities in the Lombardy region with the same industrial past that responded to the same regional call for financing urban climate strategies. After a brief introduction to the context, the first step of the analysis considers the documents presenting the climate strategies of the three administrations. The results of this content analysis are summarised in Table 10.1. In the second step, interviews and participant observation examine some aspects in depth.

Environmental history and the urbanisation of nature are characteristic elements of the three cases taken into consideration (Brescia, Bergamo and Mantova), which, before publication of the Italian NPRR, decided to apply for funding for their ecological transition at the local and European scales, starting with signing the Covenant of Majors.

Water was and remains a key element for development in Bergamo and Brescia, which exploited the energy from water and the natural elements of the nearby Alpine valleys as resources for their success as an industrial region recognised at the European and international levels. In the last years of the 20th century, Brescia began to face up to its industrial past; on the one hand, it has dealt with the pollution produced by the plants, especially but not only the metallurgical ones. On the other hand, it is redeveloping the green rural areas still present both in the pre-Alps to the north and the Po Valley to the south. Brescia also has Site of National Interest (SNI), of a large chemical industry, the former Caffaro, which has marked the city's environmental history. This has generated a series of environmental conflicts that have influenced the city's environmental policy since the 2000s. Bergamo, like the other two cities, has a long history of

Table 10.1 The three climate strategies analysed: partners, funding and main actions

Partners	*Title*	*Funding (Euros)*	*Main actions*
Municipality of Bergamo; Colli di Bergamo Park; Legambiente Lombardia (NGO); Regional Agency for Services to Agriculture and Forestry (ERSAF)	Cli.C. Bergamo!	• 5.6 m from partners • 1.3 m from the Cariplo Foundation • 1.1 m from the Lombardy region	• Actions related to energy performance linked to the reduction targets set at the European level. • Forestation interventions in the peri-urban area to improve the hydrogeological structure and reactivate cycle paths. • Depaving and restyling actions that also involve citizens. • Safely strengthening the minor water reticulum in ecological terms through naturalistic management of the riverbanks and renaturalisation of some riverbeds. • Involving citizens to identify and implement the interventions.
Municipality of Brescia; Fondazione Centre Euro-Mediterranean on Climate Change; AmbienteParco (NGO); Colline di Brescia Park	Un Filo Naturale	• 2.86 m from partners • 1.85 m from the Cariplo Foundation • 1.40 m from the Lombardy region	• Increasing natural capital and biodiversity in the urban environment. • Implementing redevelopment through depaving and rain gardens for the sustainable management of rainwater. • Creating oasis areas through planting tree and shrubs, creating green roofs on public buildings and implementing urban reforestation interventions. • Participation of citizens in some experimental design actions, involvement of stakeholders and their mapping.

Partners	*Title*	*Funding (Euros)*	*Main actions*
Municipality of Mantova; Municipalities of Curtatone, Marmirolo, Porto Mantovano, and San Giorgio Bigarello; Mincio Park; Union of Colli Mantovani; Alkémica social cooperative	ACE3T – CLIMA – Water, heat and energy: three pillars for Mantova's climate transition	• 1.80 m from partners • 0.637 m from the Cariplo Foundation • 1 m from the Lombardy region	• Energy interventions to achieve the climate neutrality targets set at the European level. • Intervention on the water reticulums of the Mincio to prevent excessive outflow of transport material that could compromise the ecosystem balance of the wetland area and on the minor reticulums to guarantee correct outflows, especially with the intensification of extreme climatic phenomena. • Forestation interventions for promoting climate adaptation and resilience in urban areas and in the rural areas of Mincio Park. • Intervention on heat waves to denaturalise and improve soil permeability in urban and suburban areas and improve hydrogeological management. • Promotion in rural contexts of ecological corridors and soft mobility along communication routes.

interactions between nature and society that have produced urban, suburban and rural landscapes. There are areas where industrial development has left clear landmarks on the urban landscape, so Bergamo is experiencing today a strong urban regeneration that is transforming these industrial areas. In contrast, in the northern part of the city, the Astino Valley, there are still urban gardens and forests that are preserved examples of old socionatures (Cisani et al., 2025; Ferlinghetti, 2019). Inside the three cities, some rural and urban areas are testimonies of socionatures produced before the urban explosion. Beginning in the 1950s, urban development erased, for the most part, these landscapes, producing the socionatures of the contemporary city. These are the old urban natures on which the new nature produced by sustainability and climate change environmental policies is grafted.

During its history, Mantova changed its surrounding natures to expand and protect the city. In particular, intervention in the lakes of Mantova transformed the landscape and interactions between the urban areas and the water of the river Mincio, producing a nature–city hybridisation that had evolved starting from the Middle Ages. The industrialisation period after the Second World War reconfigured the socionatural approach to the lakes, and their water became a resource for industry. Now, the main petrochemical sites and part of the lakes are a Site of National Interest (SNI), identified by the Italian Ministry of the Environment as an area affected by remediation activities, with consequences for health and ecosystems produced by pollutants from the plants since the 1950s.

With the full extent of the Anthropocene and the climate change era, nature is the key element of new developments in all these cities that must regreen urban spaces and renew their images as industrial cities. The nature neglected in past development nowadays is the distinct core of urban green and climate projects, and from the UPE point of view, particular attention needs to be paid to the historical and cultural paths of the socioecological interaction, the hybridity between human and non-human, and the power relations inside these green projects (Connolly, 2019; Ernstson & Swyngedouw, 2018; Tzaninis et al., 2021; Kaika et al., 2023).

The strategies adopted by the three cities (see Table 10.1) follow a common schedule in reference to the Cariplo Foundation's call for ideas, "Climate Strategy" (2021–2024) within the F2C project *Fondazione Cariplo per il Clima*.[7] The call includes three phases: first, evaluation of the proposal and the funding decision; second, drafting the climate transition strategy with the help of the foundation's experts; and third, implementation of the actions envisaged in the strategy. The initiative of the Municipality of Bergamo and its partners, "Cli.C. Bergamo!",[8] provides some actions related to water management to improve renaturalisation of the riverbanks with forestation interventions in the peri-urban area, hydrogeological risk management, reactivation of cycle paths and some depaving and restyling actions through the involvement of citizens. In particular, for the restoration of the water reticulum, action is planned to be taken in some particular peri-urban areas in the north of the city, the Astino

Valley, where the urban agricultural landscape preserves the old socionatures produced by the agricultural activities introduced by monks from the 12th century onwards.

The initiative of the Municipality of Brescia and partners, "Un Filo Naturale",[9] promotes for the city a response to climate change that makes it an "oasis city", a "sponge city" and a "city for people". Central to the strategy are eco-technical interventions, such as depaving, creating green roofs, planting trees to improve the ecological network and reforestation to increase biodiversity and improve climate adaptation and mitigation. A theme that is emphasised is the participation of citizens as well as the involvement of stakeholders in the strategy.

The Mantova Municipality and partners' initiative "ACE3T – CLIMA – Water, heat, and energy: three pillars for Mantova's climate transition"[10] aims to overcome the climatic criticalities that compromise the local hydrogeological situation with repercussions for safety (e.g., floods and heat waves), environmental quality and the production system (e.g., droughts and destructive extreme events). The first issue in the plan concerns energy to achieve the climate neutrality targets set at the European level; the second is water, in particular the river Mincio, both as a natural element to be enhanced in terms of biodiversity and as a risk factor in extreme climatic events. The third is linked to forestation interventions for promoting climate adaptation and resilience in urban and rural areas.

The contents and aims of the strategies were discussed in depth during interviews and evaluated with participant observation of some actions. The following considerations arise from these insights. In all three strategies, one of the aims is to intervene by introducing climate issues into administrative plans and instruments, to train municipal technicians through the intervention of climate transition experts and to promote citizen participation. The administrations, following the impulse responding to the same call for ideas, have been prompted to adopt some common lines of planning and intervention. All are based on alignment with certain urban policies at the European scale, especially in terms of quantifying energy production and consumption and the need to reduce climate-changing emissions, abandon fossil fuels and switch to renewable energy sources. Another common approach is illustrated by the interventions planned to ecologically improve urban contexts: depaving, reforestation and natural solutions are the practices through which to regreen cities and improve climate resilience at the European and international levels. The city of Brescia applied to the European Green Capital Award,[11] Mantova entered the European Green Pioneer of Smart Tourism Competition, and Bergamo is one of the Net Zero Cities, a European programme led by EIT Climate-KIC to reach climate neutrality by 2030.[12] These actions not only contribute to obtaining funding but also to redefining the cities' images and rebranding them through green policies to become more attractive to the outside, both nationally and internationally.

Other interesting aspects emerge from the analyses of the three cities. First, Brescia shows that citizens' participation and involvement could be a central element. Some actions of the project involve local neighbourhoods proposing and realising some climate actions based on the reintroduction of nature in urban areas. This approach has generated an interesting community awareness not only about the opportunity of the new nature-based regeneration but also about the industrial past of the city and how nature was hidden before producing some environmental conflicts.

Mantova shows a systematic approach based on its participation in different projects: an Urban Forest Forum in 2018 in collaboration with Food and Agricolture Organization, an EU Horizon project Urban Green Up and Environmental Management Audit Scheme certification. An integrated approach to different environmental and climate actions produces more diffuse benefits on related issues and in different areas of the city. Mantova's environmental actions involve diverse natures and more urban actors, spreading the interactions between ecology and society at a transcalar level. The climate strategy, from a geographical point of view, is not completely concentrated within the city limits but involves management of the natural and rural area of the Mincio river, in collaboration with Mincio Park.

Bergamo reveals particular attention to the historical hybridisation of the socionatures present in the city and to the link between the cultural approach to nature and territorial development. In particular, the experimental regeneration of part of a riverbank and restoration of the water reticulum in the Astino Valley to the north of the city demonstrate particular attention to the value of the landscape and the cultural approach to environmental management.

Urban natures between new and old socioecological interactions

This research highlights some interesting points about evolving socionatures within the studied urban areas and about the natures produced by the presented projects. First, these strategies produce and create a hierarchy of old and new natures, see Figure 10.1, that are present and interrelated today in the urban fabric of the three cases. The climate strategies of Brescia, Bergamo and Mantova produce new natures as spatial symbols of the new green urban policies that aim to demonstrate a positive interaction with nature to show a new green image of the cities and rebrand the old industrial image. Contrariwise, old natures are the spatial evidence of socionatures produced during the historical development of the three cities, including the industrial ones, that have deeply modified the urban geography of these cities and their socioecological metabolism and are not central elements of this rebranding. Future research can analyse how many different natures (old and new) will be produced in urban processes and why they are included or excluded by the climate strategies of the cities.

Climate change strategies and funding opportunities that produce new natures promote particular attention and hierarchies for some types of technical interventions, for example, depaving, reforestation, the creation of oasis

Figure 10.1 Examples of new and old urban socionatures in the case studies. Photographs by the author.

areas within the city or photovoltaic installations in rural areas. However, a critical approach needs to be taken to understand whether these interventions are chosen only because of funding opportunities and whether other socioecological problems need to be addressed first. The risks are that only some specific parts of the cities, important for their centrality, benefit from this intervention because the experimental project needs to be visible, as in the case of the Astino Valley in Bergamo or the participatory actions in Brescia placed in the city centre, while some other more peripheral areas of the city with socioecological problems do not have the opportunity to be included in the project.

Other complications relate to the experimental approach of these urban climate strategies. They do not automatically produce a cultural change in the logic of management of urban natures. As emerged from the interviews, the experts, municipal technicians and managers directly involved in the projects show a strong motivation to produce socioecological transitions and a real understanding of the nature–urban relationship. However, offices of the administration not involved in the project, private experts or specialists and economic actors need to be involved in this cultural change. All three projects produce a climate strategy as part of their outputs, but it is necessary to verify whether in the future this will effectively influence environmental urban policies and plans and spread the attention to nature to all urban stakeholders or remain a simple experimentation to finance some location-specific reproductions of nature.

Research on urban nature starts from an understanding of the urban socionatures present in cities to recognise different positive, negative, ecological, cultural and social aspects. A project or action based on natural elements is not good in itself. For this, an atlas of these socionatures interesting urban areas must be the starting point to produce and integrate greening policies based not only on the production of supposed natural elements within the urban project but also on valorisation of the historical, cultural and socioecological peculiarity of each city.

A preliminary study of the socionatures could reveal conflicts and socioecological problems within the city, without forgetting the other socionatures considered less positive and characterised by complications and contradictions, but still part of the development of the socioecological system. In the future, it could be useful to promote a local network between the three cities, with the aim of achieving positive results through exchanges about the strengths and weaknesses of each approach and experiences of mapping the types of urban socionatures.

Notes

1 https://population.un.org/wup/Download/Files/WUP2018-F17a-City_Size_Class.xls
2 https://commission.europa.eu/strategy-and-policy/priorities-2019-2024/european-green-deal_en
3 https://eu-mayors.ec.europa.eu/en/home
4 https://environment.ec.europa.eu/topics/urban-environment.european-green-capital-award_en

5 https://www.mase.gov.it/pagina/piano-la-transizione-ecologica
6 https://www.eea.europa.eu/themes/air/urban-air-quality/european-city-air-quality-viewer
7 For more information on the call and the Fondazione Cariplo's climate initiatives, see the webpage at: https://www.fondazionecariplo.it/it/progetti/ambiente/f2c-fondazione-cariplo-per-il-clima.html
8 https://www.comune.bergamo.it/node/366480
9 A Natural Tread, https://www.comune.brescia.it/aree-tematiche/urban-center/progetto-un-filo-naturale/un-filo-naturale-una-comunita-che-partecipa/un-filo-naturale-una-comunita-che-partecipa
10 https://www.comune.mantova.it/index.php/clima/2115-ace-3t
11 https://environment.ec.europa.eu/topics/urban-environment/european-green-capital-award_en
12 https://netzerocities.eu/

References

Angelo, H. (2020). *How green became good, urbanized nature and the making of cities and citizens*. University of Chicago Press.
Beatley, T. (2012). *Green cities of Europe*. Island Press.
Braun, B. (2005). Environmental issues: Writing a more-than-human urban geography. *Progress in Human Geography*, *29*(5), 635–650. https://doi.org/10.1191/0309132505ph574pr
Brenner, N. (Ed.). (2014). *Implosions/explosions. Towards a study of planetary urbanization*. Jovis.
Broto, V. C., & Bulkeley, H. (2013). Maintaining climate change experiments: Urban political ecology and the everyday reconfiguration of urban infrastructure. *International Journal of Urban and Regional Research*, *37*(6), 1934–1948. https://doi.org/10.1111/1468-2427.12050
Bryant, R. L., & Bailey, S. (1997). *Third world political ecology*. Routledge.
Caprotti, F., & Cowley, R. (2017). Interrogating urban experiments. *Urban Geography*, *38*(9), 1441–1450. https://doi.org/10.1080/02723638.2016.1265870
Caprotti, F., & Yu, L. (Eds.) (2017). *Sustainable cities in Asia*. Routledge.
Castree, N., & Braun, B. (2001). *Social nature: Theory, practice, and politics*. Blackwell Publishers.
Cisani, M., Tononi, M., & Ferlinghetti, R. (2025). Peri/urban and regional protected areas. Structure and prospects of the sociopolitical nature of conservation in Lombardy. In S. Bonati, M. Cisani, M. Tononi, & G. Zanolin (Eds.), *Italian natures. Perspective on the social nature in the "Bel Paese"*. Routledge.
Connolly, C. (2019). Urban political ecology beyond methodological cityism. *International Journal of Urban and Regional Research*, *43*(1), 63–75. https://doi.org/10.1111/1468-2427.12710
Desfor, G., & Keil, R. (2004). *Nature and the city: Making environmental policy in Toronto and Los Angeles*. University of Arizona Press.
Ernstson, H., & Swyngedouw, E. (Eds.) (2018). *Urban political ecology in the Anthropo-obscene: Interruptions and possibilities*. Routledge.
Evans, J., Karvonen, A., & Raven, R. (2017). *The experimental city*. Routledge.
Ferlinghetti, R. (2019). Processi di risignificazione e rigenerazione della corona dei Corpi Santi di Bergamo. In B. Castiglioni, & S. Zaggia (Eds.), *Monastero e territorio: periferie dello spirito e dello spazio* (pp. 69–84). Padova University.
Forsyth, T. (2003). *Critical political ecology*. Routledge.
Gandy, M. (2005). Cyborg urbanization: Complexity and monstrosity in the contemporary city. *International Journal of Urban and Regional Research*, *29*(1), 26–49. https://doi.org/10.1111/j.1468-2427.2005.00568.x

Gandy, M. (2022a). Urban political ecology: A critical reconfiguration. *Progress in Human Geography*, *46*(1), 21–43. https://doi.org/10.1177/03091325211040553
Gandy, M. (2022b). *Natura Urbana: Ecological constellations in urban space*. The MIT Press.
Harvey, D. (1993). The nature of environment: Dialectics of social and environmental change. *Social Register. Real Problems, False Solutions*, *29*, 1–51.
Häußler, S., & Haupt, W. (2021). Climate change adaptation networks for small and medium-sized cities. *SN Social Sciences*, *1*(11), 262. https://doi.org/10.1007/s43545-021-00267-7
Heynen, N., Kaika, M., & Swyngedouw, E. (2006). *In the nature of cities: Urban political ecology and the politics of urban metabolism*. Routledge.
Kaika, M., Tzaninis, Y., Mandler, T., & Keil, R. (2023). *Turning up the heat: Urban political ecology for a climate emergency*. Manchester University Press.
Lefebvre, H. (1970) *La révolution urbaine*. Gallimard.
Negreiros, P., & Falconer, A. (2021). Financing the green transition of european cities: What does the european green deal change. In Hannah, A. (Ed.), *Towards a European green deal with cities. the urban dimension of the EU's sustainable growth strategy* (pp. 49–59). CIDOB. https://www.cidob.org/sites/default/files/2024-07/GREEN%20DEAL.pdf
Robbins, P. (2004). *Political ecology: A critical introduction*. Blackwell.
Smith, N. (1984). *Uneven development*. Blackwell.
Swyngedouw, E. (1996). The city as a hybrid: On nature, society and cyborg urbanization. *Capitalism Nature Socialism*, *7*(2), 65–80. https://doi.org/10.1080/10455759609358679
Tzaninis, Y., Mandler, T., Kaika, M., & Keil, R. (2021). Moving urban political ecology beyond the 'urbanization of nature'. *Progress in Human Geography*, *45*(2), 229–252. https://doi.org/10.1177/0309132520903350
Whatmore, S. (2002). *Hybrid geographies: Natures cultures spaces*. Sage.
Wheeler, S. M., & Beatley, T. (2004). *The sustainable urban development reader*. Routledge.

11 Greening the "void"

Brownfield redevelopment and soil desealing in two Italian cities

Emanuele Garda

Introduction

The growth of cities is not spurred by their own instincts; rather, they are built piece by piece by their inhabitants (Rykwert, 1976). Communities continuously modify their host cities through deliberate actions, expanding them in multiple dimensions according to an alternation between use, disuse, and reuse that appears to be a fundamentally physiological constant in the history of human societies (Scaramellini, 2018). This latter hypothesis helps us situate the phenomenon of the decommissioning of industrial areas, which first took place in the 1980s in Europe and North America and triggered a long phase of applications. This was a complex season that recognised in the recovery of brownfields a common challenge for public policies (Atkinson et al., 2014).

In the debate that began in the 1980s, reference was often made to the concept of an "urban void" in descriptions of industrial areas temporarily abandoned due to decommissioning, with this absence characterised only in terms of a lack of use of such areas by humans. This interpretation did not consider the opportunities that human neglect provides for flora and fauna, which, with human abandonment, found greater freedom to progressively recolonise this vacant land (Németh & Langhorst, 2014). In this transition from a society with an industrial identity to one with a new post–industrial character (Arcidiacono et al., 2015), different methods of reconceiving these areas have emerged. In addition to making existing structures once again functional and constructing new buildings by means of demolition and reconstruction, one can find accounts of different approaches. For instance, it is possible to identify processes that have implemented partial demolition of built-up areas and effectuated a consequent increase in soil permeability and renaturalisation.[1] With this third approach, which is oriented towards the idea of an "urbanism of demolition" (Rusci, 2021), it is possible to liberate previously "sealed" areas, thereby creating the conditions for the restoration of certain features inhibited by sealing processes (Garda, 2020) and the promotion of greening. This idea of the removal of built-up spaces is captured in the concept of desealing (Garda, 2022), which can be defined as "restoring part of the former soil profile by removing sealing layers such as asphalt or concrete" (EU, 2012, n.p.).

DOI: 10.4324/9781003465164-15

The removal of built-up surfaces and structures is seen as a necessary condition for freeing soils from artificial constraints and providing space for renaturalisation and the implementation of nature-based solutions. A similar interpretation is suggested by Tobias et al. (2018), for whom "soil unsealing", achieved through the removal of artificial surfaces, represents a necessary act of restoring areas to a state of (almost) "non–land use", which implies the reduction of the restricted use of areas for human needs alone. The authors assert that it is possible to foster the formation of urban parks, encourage the return of naturalness with pioneer species, or even allocate these spaces to agriculture or forestry. The advantages that can be obtained by these interventions have great importance for urban areas, especially those characterised by intense soil sealing processes (Garda, 2020) due to the negative effects produced by anthropogenetic soil sealing (Scalenghe & Marsan, 2007; EEA, 2013).

Since the post-World War II period, Italian cities have been affected by a massive soil urbanisation cycle that has entailed the loss of substantial amounts of natural space. In this extraordinary and dramatic season (Lanzani & Pasqui, 2011), urban growth has impacted the spaces identified by local communities, through their urban plans, as expendable to meet the needs of demographic and economic growth to the detriment of the system of green areas present inside, and especially outside, the cities. Communities in the grip of the so-called "concrete fever" (Calvino, 1994) have sacrificed their relationship with nature to the drive to acquire developable land sealed with a "permanent plaque" that has rendered it inert. This has profound effects on the "four ecospheres" (the geosphere, hydrosphere, atmosphere, and biosphere) and alters ecosystemic functions within the carbon and water nutrient cycles (Pavia, 2015).

Within this critical framework, desealing emerges as a compensatory action (EU, 2012) that is consciously adopted to remedy past mistakes, which are even more evident today in association with the negative effects of climate change. Therefore, through desealing, it is possible to (i) increase green areas, which are crucial for the health of inhabitants (Kabish et al., 2015); (ii) reduce urban heat islands (Magliocco & Perini, 2014); (iii) improve water cycle management (Brears, 2018); (iv) promote ecological reconciliation (Francis & Lorimer 2011); and (v) (re)activate ecosystem functions (Garda et al., 2023).

The concept of desealing also has another meaning. When applied to already built-up spaces, these actions are confronted with the obsolescence of these places and the opportunity to foresee their demolition. In our new season, in which the recycling paradigm (Ciorra & Marini, 2011) is seen as the main strategy, it is necessary to assume that "not everything can be recycled" (Merlini, 2019). Therefore, assuming that parts of the buildings present in the Italian territory are to be understood as "unrecyclable" (Lanzani et al., 2014) supports "the path of demolition" (Merlini, 2019). Cities are a social product built and shaped over time (Secchi, 2013), and the need to intervene to maintain the entire "community house" (Salzano, 1998) is not new or unusual. Cities continue to express a "need to be repaired" (Sennet, 2018) to respond to new needs or, as noted above, to remedy past mistakes. At the same time, the city must be considered a socioecological system (Pietta & Tononi, 2021) in which society and nature interact, thereby determining influences on and changes in urban metabolism (Heynen et al., 2006). This strong connection between community and nature is familiar to urban planners. Between the 19th and 20th centuries, many authors[2] formulated hypotheses about urban models centred on the presence of green areas within built spaces, recognising their redeeming role in promoting the health of residents.

This paper addresses the issues described above through an analysis of two cases in the Italian cities of Trento and Milan. In both cases, we are dealing with complex processes of radical transformation in areas that were previously completely built-up and lacking in natural elements or relations with nature. Transformational interventions have favoured the creation of new green areas through masterplans that have given a central role to partial renaturalisation and the need to re-establish a relationship between communities and nature, considering these design strategies to be characteristic priorities.

The two case studies were selected because of certain common features:

- All the interventions were almost fully realised.
- The industrial areas were homogeneous in terms of architectural and ownership structures.
- Before the transformation, the areas had high levels of soil sealing and a total absence of natural elements.
- The projects that guided the transformation reconfigured the two areas.
- The creation of new green areas was a core aspect of the planning and design tools employed.

Therefore, these were top–down planning and design processes that, for this reason, inspired the greatest direct participation among public institutions and private stakeholders directly affected by the transformation (with the limited involvement of the local community).

Le Albere district

The history of the former Michelin facility in Trento began in the 1920s, when the company decided to build a production plant in the area. The importance of this plant for the city was not only due to its territorial extension (116,000 sqm) but was also supported by other reasons. A significant role was played by this economic operator, who owned this site for decades, shaping the labour market (with 1,800 workers in the 1970s) and community life. Even the functional choices proposed and then realised during this transformation process influenced the new district because of new territorial relations as well as the interaction with local territorial elements (e.g., the Adige River). At the same time, the Michelin plant can be considered a metaphor for a city that has been confronted with structural phenomena (Mazzeo, 2016), such as the end of a long cycle of demographic growth and the crisis of the industrial sector. Following

Marcello Vittorini's urban plan (1989), the city cycled through many modifications to the plan in an attempt to address the needs of an area in search of a new balance.

The district that replaced the decommissioned industrial site, now known as Le Albere, is the outcome of a process that, since 1998 – the year in which the Michelin company decided to shut down operations – has solicited stakeholders, places, and planning tools. The first proposals for the area had already been introduced in the 1989 urban plan, which envisaged a fragmentation of the structure into mixed–residential blocks set in green spaces. In response to local criticism, this plan was abandoned in favour of a variant of the urban plan in 1994 (Ulrici, 2011). A few years later (1998), the City Council returned to this issue, initiating a new revision of the municipal plan.

A 2001 update to the municipal plan played a fundamental role in this new trajectory. The first masterplan was drafted in 2003 by the Renzo Piano Building Workshop, which had been commissioned by the new owner of the area (the *Iniziative Urbane*). The involvement of this important architect not only lent the area's redevelopment project international importance; it also defined and branded the entire transformation process in terms of both the overall structure of the new district, including the new green areas, and the individual buildings constructed (also designed by Renzo Piano).

The two main design choices proposed by the project concerned the creation of a multiscalar mix of functions. The second theme dealt with the choice to concentrate the new building volumes in a limited part of the area, creating the foundational conditions for a large urban park. The masterplan provided for a series of interventions along the boundaries of the project area, which were fundamental in accelerating the realisation of a more comprehensive urban redevelopment of the district (Antoniaconi, 2010). These interventions concerned the construction of several pedestrian subways under the (elevated) railway, road, and pedestrian bridges over the Adige, and the construction of interchange car parks. These efforts also included a project of particular importance for the new park and for integration with the Adige River: the burial of a public street (Via Sanseverino) to instal a clear barrier between the area and the river. By burying this road, it was possible to create a unitary park that connects the heart of the new district with the environmental corridor of the Adige River, leaving no visual or fruitive discontinuities.

After the City Council's approval of the masterplan (2004), the subdivision plan's turn came the following year. Responsibility for drafting was once again assigned to Renzo Piano's Studio, guaranteeing a certain continuity in the design approach. The importance attributed to the green system and, above all, the proposal to create a large park area were again recognised as fundamental to the new district.

The construction work for the new district, later renamed Le Albere, began in 2008 and led to the construction of new structures for residential (44,000 sqm), tertiary (29,000 sqm), and commercial (10,500 sqm) use. Another important and symbolic intervention concerned the construction of a Science Museum, again designed by Renzo Piano, located in the northern part of the district. In addition to an important share of mainly underground parking areas (2,000 parking spaces), the interventions in the open spaces must also be mentioned. They included squares, pedestrian paths, and water channels (about 28,000 sqm) as well as the construction of a public park, which affected a large portion of the district (50,000 sqm).

OM–Pompeo Leoni district

The history of the former OM of Milan, an area located to the south in a particularly urbanised setting, is rather extensive. This history began at the end of the 19th century when the company Miani Silvestri, then engaged in the manufacture of railway materials, decided to settle in this part of the city. A few years later (1899), the area took on the name *Officine Meccaniche* (OM). In a short time, thanks to the effects of its proximity to the Porta Romana railway station, it became one of Milan's most important industrial centres (4,000 workers). Starting in the 1930s, this company underwent various changes in management and ownership structure, gradually becoming part of the FIAT Group. Between the 1980s and 1980s, the divestment of production activities began, marking a historical moment that, for Milan, represented the start of a powerful season of reflection on abandoned areas.

From the point of view of urban policies, the debate over the fate of the large "urban voids" in Milan was tackled by means of newly planned instruments that differed radically from the more traditional town-planning schemes. A first act produced by the municipality, introduced as a strategic tool for coordinating public and private interests (Bolocan, 2009), was the *Documento direttore per il progetto passante* (Oliva, 2002). Parallel to the first document was another tool (*Studi di inquadramento*) that, by way of a more in-depth analysis of Milan's underused areas, suggested new strategies. For OM, the theme and potential role of urban greenery emerged for the first time, suggesting the need to create a "green system" capable of redefining the relations between the city and its peripheral agricultural spaces (Doni, 2009).

In the mid-1990s, a second important moment for urban policies unfolded in the *Nove Parchi per Milano* (1995) project. The purpose was to define new centralities and create a large urban park (Marinoni, 2007) in the peripheral areas that would contrast with the monocentrism of the *forma urbis* of Milan.

The transformation of the area began with the municipality's participation in a national funding effort intended to support urban regeneration. After identifying the areas to be evaluated for a possible request for financing, real estate companies submitted a proposal for the *Programma di riqualificazione urbana* (PRU) of the former OM, later named PRU OM–Pompeo Leoni, for a total area of 313,652 square metres, which was subsequently approved by the Municipality and the Ministry. The Programme Agreement was signed in 1997 by the Municipality, the Lombardy Region, and the Ministry of Public Works. Work began in 1999, while construction was mainly concentrated between

2003 and 2005. The reconversion of the area led to the creation of a new neighbourhood characterised by a rather ordinary mix of functions. On the whole, of the approximately 153,000 sqm of floor area ratio, a very significant portion (79,000 sqm) was set aside for residential use following the strong demand for housing in the local market. The remaining areas were reserved for commercial structures (9,000 sqm), a tertiary sector (34,000 sqm), and manufacturing functions (31,000 sqm).

Comparing the two projects

Of the approximately 116,000 sqm allocated for the Michelin plant project, a considerable portion was reserved for new green areas (43%), which represented a strategic decision in the redevelopment process. The creation of these green areas was made possible by a combination of the complete demolition of existing structures and the concentration of building potential in the eastern part of the district. The increased permeability of the areas manifested as the result of the project of converting an area that, due to the morphological and functional conditions of the previous activities, was characterised by a high degree of soil impermeability. Desealing became necessary to facilitate the creation of a new public park and meet certain needs (Garda, 2020). The first need concerned the attractive role given to the large, equipped park, which was supported by the provision of specific elements. The spatial components intended to ensure attractiveness (for new inhabitants and other non-residents) and define the park's spatial composition included the creation of paths, equipped meadows, and wetlands. The second priority was to reconnect with the Adige River by encouraging a return to the dialectical relationship that had been disrupted by urbanisation. Likewise, the park was considered a link between the city and the river.

The proposed design framework adopted a taxonomic approach because the different "elements" that made up the large, equipped park are treated separately within that framework. In the technical documentation, the spaces were organised through the use of "diversified landscape environments", which comprised (i) a central environment treated as a lawn and equipped with paths, (ii) a wet garden composed of species typical of river environments, and (iii) groves consisting of prime tree species distributed along the edges.

The system of green areas that now characterises Le Albere faithfully recreates the design directions specified by the transformation project. It is possible to recognise both the spatial design imagined in the various elaborations of the project and the variety of elements that characterise it, which represent a certain complexity by facilitating a return of nature within this urban area.

The principles adopted for the former OM area were centred on a different approach. According to the design, proposed by French landscape architect Frederic Girot, the new park was organised into different thematic areas. This proposal matched the instructions for the project carried out by the municipal administration, albeit with some limitations related to the impossibility of surmounting the physical barrier constituted by the railway line. The obvious obstacles prevented the implementation of the idea of a unified public park, which was also envisaged by the *Nove Parchi per Milano* project and achieved via the connection between the existing Ravizza Park, located to the north, and the new park planned for the former OM. The architect proposed a park divided into three environments alternating with the new buildings, organised from east to west, and integrated with the Vettabbia Canal, which ran through the entire area (and for which reclamation was planned).

Further north, between the railway and Viale Toscana, a Culture Park was planned. It was characterised by the construction of outdoor sports facilities, a large amphitheatre for 1,500 people, and new wooded and planted areas with fruit trees. Planned for the centre was the Park of Industrial Memories, which was intended to facilitate integration between new green areas (lawns and rare trees), bicycle and pedestrian paths, and certain technological and architectural elements maintained to preserve the historical memory of the area. To the south, the project envisioned Vettabbia Park, which derived its name from the presence of the Vettabbia Canal. In this case, the design indications were oriented towards enhancing the canal through the creation of bicycle and pedestrian paths, tree-lined strips, and small squares identified with the new lawn areas to be placed between the Vettabbia Canal and the newly built-up areas. Between Memorial Park and Vettabbia Park were the areas reserved for the construction of new residential and tertiary buildings.

At the centre of the built-up spaces, from east to west, ran the main road axis for the entire project, whose design was characterised by the presence of paths alternating with green spaces (lawns and tree rows) designed to improve the environmental quality of the new roads. In its current condition, this part of the new neighbourhood can be considered a boulevard and a distinctive thematic section.

The various thematic areas were connected along the north–south axis by pedestrian/cycle paths intended to provide continuity between these areas due to the discontinuities projected to interrupt the otherwise wide frontage of new blocks and buildings. Girot's proposals were later integrated into a project carried out in 1998–1999 by the landscape architect Andreas Kipar from LAND Studio, who developed the executive proposal by adapting it to new requirements agreed upon with the City of Milan without radically altering the original project. Park construction was started immediately and completed in 2006, except for Culture Park, which is currently still being concretised following the decision, after several decades of disuse and the formation of large fragments of "third landscapes", to build an underground station for electric buses and new green-roofed areas set aside for leisure-related functions.

Discussion

The two projects reviewed above are characterised by a common interest in the return of nature, mainly in the form of an urban park, to areas previously filled by impervious surfaces and industrial buildings. These case studies allow us to reflect on some general themes. The first concerns the concept of desealing, which, in these two processes, found full implementation through the demolition of buildings and impervious surfaces, an increase in soil permeability, and the creation of green areas. In both cases, the new permeability was the indirect result of design approaches primarily interested in creating a new ecology or supporting the demand for new recreational spaces.

This theme is connected to the issue of the language that guided the two proposals, wherein concepts were used that are not completely consistent with the current debates on urban regeneration and climate change, or with other increasingly common terms (e.g., NbS, ecosystem services, green infrastructure). Despite this difference – if only in the concept of desealing, which is almost absent – the new neighbourhoods make an active contribution to the urban contexts in which they are embedded, for example, in terms of the ecosystem services provided (for microclimate regulation and temperature reduction).

Another theme concerns the planning tools used to guide the transformation. While beginning with the same starting conditions, that is areas with similar morphological characteristics, the cases involve the adoption of different planning and design tools. For the former Michelin plant, the redevelopment process was sustained through a traditional approach, both with respect to the direct relationship between the directions in the municipal plan and the subsequent implementation plan. In the former OM, a different path was taken that benefited from instruments introduced at the local scale directly by the municipality, and other instruments introduced at the national level. Given the similar results, at least in terms of a high allocation of vacant areas, it could be argued that the adoption of new instruments is not always essential for implementing an innovative idea.[3]

A comparison of the structures and composition of the new green spaces reveals differences in spatial organisation. In Trento, the park and the network composed of the other secondary elements have the function of ensuring integration between the various spaces, but above all, they create strong relationships with an important ecological element lying beyond the boundaries of the new neighbourhood. The concentration of the new green spaces, especially of the more forested areas, along the west side, and the burial of Via Sanseverino are part of a strategy to redefine the relationship between the city and the Adige River. There is considerable variety and diversity among spaces intended to facilitate outdoor recreational functions, forested areas, and wetlands. This variety not only ensures a certain diversity in ecosystemic functions but also educates and disseminates information consistent with what is presented at the Science Museum in Le Albere. These are spaces between two-dimensional (meadows) and three-dimensional elements (rows or forested areas) that are framed using vegetation essences consistent with the territorial setting.

The general structure of the green system present in the former OM has different features. Its division into thematic areas and the absence of prominent structural and multiscalar elements, as in the case of Trento, weaken the perception of a general pattern of organisation. When observing these spaces, which are mostly owned by the municipal administration, one gets the sense of a buffer zone positioned to separate (Bricocoli & Savoldi, 2010) the new residential architecture built at the centre of the neighbourhood from the urban surroundings, promoting a spatial, symbolic, and psychological boundary. From a compositional point of view, the new surfaces come across as less structured due to the greater presence of lawns and trees, which, on the whole, correspond more to the recreational goals and residential functions of the neighbourhood and the urban context. From an ecological point of view, therefore, the case of Milan exemplifies less variety and biodiversity interest than the case of Trento.

Both projects promote the idea of a "reconciliation" between humans and nature by removing artificial structures in favour of a return of natural areas. Of course, such efforts cannot completely restore what has been lost through urbanisation and can only plant the seeds of a new beginning. In both cases, the new green areas were the result of very detailed planning in terms of both the spatial distribution of the natural components and their typological characterisation. They are thus not the result of spontaneous renaturalisation, like that which has affected abandoned areas over the decades, but rather of an approach to the construction of green areas similar to that used for the design of built-up areas. In the designers' intentions, however, an attempt emerged to bring nature back into the lives of inhabitants, workers, and users of collective functions, whether through the new architecture or in the urban context of reference, through the creation of green areas intended to support human functions and practices. In the two projects and their realisations, we recognise an attempt at a dialogue with pre-existing natural elements, for example in the plantings that proposed types typical of the area. In the proposed designs, however, these elements took on the appearance of an artificial environment.

Conclusions

Denaturalisation interventions play a key role in the transformation and greening of large urban voids by fostering the improvement of built spaces (Lehmann, 2021), which are valued for their ability to remind people of the importance of biodiversity, ecosystem services, and green infrastructure.

The two cases examined offer an opportunity to observe how topical issues and needs – for example, increasing soil permeability and promoting renaturalisation interventions to improve the urban environment and counteract the problems generated by soil sealing and climate change – can also find "unconscious" support within urban design processes that arose at a stage that anticipated the full awareness we now have of the many global crises (Garda, 2022).

This unawareness concerns the adoption of design approaches and concrete interventions that, while supported by language seemingly distant from the concepts adopted today, we can nonetheless reinterpret with today's sensitivity and awareness, thanks above all to the actual outcomes of such choices.

This also supports the idea that, to implement innovative interventions consistent with current needs, we should not necessarily wait for the introduction of new technical and regulatory tools but, on the contrary, seek to reorient even the most "traditional" methods. The redesign of urban voids and concomitant release of soil are the first steps in a series of possible shifts towards the return of green areas to dense urbanised areas after a long season of growth.

The opportunity provided by soil desealing must be understood only in terms of performance, for example, as a function of increased hydraulic functionality, or economically, as a factor in the valorisation of newly constructed private buildings (assuming, in this case, a form of commodification). Rather, they must be seen as an opportunity to bring humans and nature closer together, thereby diminishing or annihilating the dichotomy produced by 20th-century planning.

Notes

1 This concept includes all planned and designed interventions, as opposed to spontaneous processes attributable to the concept of the third landscape (Clément, 2004). The latter concerns the creation of new green areas endowed with vegetation elements of various types that are explicitly designed to be used by human populations for specific practices (e.g. walking, sports, or recreation). These interventions are primarily concerned with the shaping of a nature intended to respond to community needs through the creation of well–defined spaces.

2 One could mention the proposals of socialist utopians such as Robert Owen in the early 19th century, or the plan realised in the middle of that century by Ildefonso Cerdà for Barcelona, which envisaged the creation of blocks encompassing 65% green areas. With respect to the 20th century, numerous city models can be mentioned, such as Ebenezer Howard's "Garden City", which became famous precisely for its interaction and placement of settlements within a green belt, or Le Corbusier's "Ville Radieuse" project, which envisaged gigantic skyscrapers within large continuous green areas.

3 The Italian legislative framework on town planning provides for a multiscalar model of the transformation of an area of special significance, which consists of the adoption of an implementation instrument prior to individual architectural projects. At the urban scale, this implementation tool traditionally had to respect the rules dictated in the municipal plan.

References

Antoniaconi, G. (2010). L'amministrazione alla prova. In G. Codecasa (Ed.) *Governare il partenariato pubblico e privato nei progetti urbani*. Maggioli Editore.

Arcidiacono, A., Bruzzese, A., Gaeta, L., & Pogliani, L. (2015). *Governare i territori della dismissione in Lombardia*. Maggioli Editore.

Atkinson, G., Doick, K. J., Burningham, K., & France, C. (2014). Brownfield regeneration to greenspace. *Urban Forestry & Urban Greening*, *13*, 586–594. https://doi.org/10.1016/j.ufug.2013.04.002

Bolocan, M. (2009), *Geografie milanesi*, Politecnica.

Brears, R. C. (2018). *Blue and green cities. The role of blue–green infrastructure in managing urban water resources*. Palgrave Macmillan.

Bricocoli, M., & Savoldi, P. (2010). *Milano downtown*. Edizioni.

Calvino I. (1994). *Le città invisibili*. Mondadori, Milano.

Ciorra, P., & Marini, S. (Eds.). (2011). *Re-Cycle. Strategie per l'architettura, la città e il pianeta*. Electa.

Clément, G. (2004). *Manifeste du tiers paysage*. Sens & Tonka.

Comune di Milano. (1995). *Nove Parchi per Milano*. Electa, Milano.

Doni, M. (2009). Gli effetti urbanistici di un grande progetto di trasformazione della città di Milano. *Territorio*, *47*, 139–149. https://doi.org/10.3280/TR2008-047016

EEA. (2013). *Hard surfaces, hidden costs*. Office for Official Publications of the European Communities.

EU. (2012). *Guidelines on best practice to limit, mitigate or compensate soil sealing*. Publications Office of the European Union.

Francis, R. A., & Lorimer, J. (2011). Urban reconciliation ecology: The potential of living roofs and walls. *Journal of Environmental Management*, *92*,1429–1437. https://doi.org/10.1016/j.jenvman.2011.01.012

Garda, E. (2020). *I territori delle densità. Letture e interpretazioni tra crescita e contrazione urbana*. Aracne.

Garda, E. (2022). Reasons, concepts, and methods for soil de–sealing in the regulatory framework of the Italian regions. *IJPP*, *12*(1), 65–81.

Garda, E., Marucci, A., & Falasca, F. (2023). De–sealing, recupero del suolo e (ri) attivazione dei Servizi Ecosistemici. *TRIA*, 107–126. https://doi.org/10.6093/2281-4574/10600

Heynen, N., Kaika, M., & Swyngedouw, E. (2006). *In the nature of cities: Urban political ecology and the politics of urban metabolism*. Taylor & Francis.

Kabish, N., Qureshi, S., & Haase, D. (2015). Human–environment interactions in urban green spaces. *Environmental Impact Assessment Review*, 50, 25–34. https://doi.org/10.1016/j.eiar.2014.08.007

Lanzani, A., Merlini, C., & Zanfi, F. (2014). Quando «un nuovo ciclo di vita» non si dà. *Archivio di Studi Urbani e Regionali*, 109, 28–47. https://doi.org/10.3280/ASUR2014-109003

Lanzani, A., & Pasqui, G. (2011). *L'Italia al futuro. Città e paesaggi, economie e società*. Franco Angeli.

Lehmann, S. (2021). Growing biodiverse urban futures. *Sustainability*, *13*(5), 2932. https://doi.org/10.3390/su13052932

Magliocco, A, & Perini, K. (2014). Urban environment and vegetation: Comfort and urban heat island mitigation. *Techne*, *8*, 155–162.

Marinoni, G. (2007). Milano. Città in evoluzione. *LOTUS*, *131*, 132–141.

Mazzeo, G. (2016). *La città leggera*, Federico II University Press.

Merlini, C. (2019). L'eventualità della demolizione. *Archivio di Studi Urbani e Regionali*, *124*, 26–48. https://doi.org/10.3280/ASUR2019-124002

Németh, J., & Langhorst, J. (2014). Rethinking urban transformation: Temporary uses for vacant land. *Cities*, *40*,143–150. https://doi.org/10.1016/j.cities.2013.04.007

Oliva, F. (2002). *L'urbanistica di Milano*. Hoepli.

Pavia, R. (2015). *Il passo della città: temi per la metropoli futura*. Donzelli.

Pietta, A., & Tononi, M. (2021). Re–naturing the city: Linking urban political ecology and cultural ecosystem services. *Sustainability*, *13*, 1786. https://doi.org/10.3390/su13041786

Rusci, S. (2021). *La città senza valore*. Franco Angeli.

Rykwert, J. (1976). *Idea of a town*. Princeton University Press.

Salzano, E. (1998). *Fondamenti di urbanistica: la storia e la norma*. Laterza.

Scalenghe, R., & Marsan, F. A. (2007). The anthropogenic sealing of soils in urban areas. *Landscape and Urban Planning*, *90*, 1–10. https://doi.org/10.1016/j.landurbplan.2008.10.011

Scaramellini, G. (2018). Uso disuso riuso. fisiologia o patologia del territorio?. In F. Adobati, & E. Garda (Eds.), *Biografie sospese* (pp. 37–57). Mimesis.

Secchi, B. (2013). *La città dei ricchi e la città dei poveri*. Laterza.
Sennet, R. (2018). *Costruire e abitare: Etica per la città*. Feltrinelli.
Tobias, S., Conen, F., Duss, A., Wenzel, L., Buser, C., & Alewell, C. (2018). Soil sealing and unsealing: State of the art and examples. *Land Degrad*, *29*, 2015–2024. https://doi.org/10.1002/ldr.2919
Ulrici, G. (2011). La riqualificazione dell'ex Michelin: la città e l'archistar. *Urbanistica informazioni*, *237*, 53.

12 (Peri)urban and regional protected areas

Structure and prospects of the sociopolitical nature of conservation in Lombardy

Margherita Cisani, Marco Tononi, and Renato Ferlinghetti

Urbanisation and nature conservation

Nature conservation has historically been an issue restricted to non-urban spaces, linked to the experience of nature by a bourgeois class inspired by Romanticism (Cronon, 1996). Nevertheless, "urbanisation and Romantic enthusiasm for sublime and wild nature share an intertwined history" (Aschenbrand, 2024, p. 4). The emergence of national parks shares some characteristics with the development of public urban parks, especially those planned and promoted by the landscape architect and public administrator Frederick Law Olmsted, such as Central Park in New York, preserved for the collective enjoyment of "nature" in an urban environment (Carr, 2000). Moreover, the urban and dichotomous view of nature as places considered wild and untouched that is characteristic of mainstream conservation is gradually moving closer to people, seeking greater affective connection and stewardship and addressing spaces near, or even within, densely urbanised areas.

This process is closely linked to the ongoing changes in the human–nature relationship in the Anthropocene (Büscher & Fletcher, 2020; Lorimer, 2015). On the one hand, the need for nature is growing and is increasingly recognised – for example, through ecosystem services and nature's contributions to people but also through growing ecotourism practices. On the other hand, it is increasingly acknowledged that this nature is produced and reproduced due to sociopolitical dynamics that have particularly affected, albeit in different ways, peri-urban contexts. We specifically refer to the so-called third landscape – that is, neglected areas often found on the outskirts of cities (Clément, 2004) – "minimal landscapes" (Ferlinghetti, 2011), agricultural land or brownfield rewilding (Piana et al., 2024), extractive industry sites transformed into natural habitats (Pietta & Tononi, 2021) and rural ecosystems that have historically evolved in relation to the city (Scazzosi, 2020). In such contexts, naturalness finds unprecedented spaces for conservation and evolution, so much so that a new term, *urban wildness* (Loi & Pierucci, 2025; Metta & Olivetti, 2019; Müller et al., 2018), has been introduced to refer to the species and biocoenoses that inhabit them.

DOI: 10.4324/9781003465164-16

Following the social nature approach (Castree & Braun, 2001), the study of the development of protected areas goes beyond the dichotomies between society and nature and between city and nature. According to this approach, protected areas can be considered the result of the interrelationships between social and ecological dynamics that produce particular socionatures. According to Swyngedouw (1996, p. 3), "Nature itself is a historical-geographical process", he insists on the inseparability of society and nature, mantaining "the unity of socio-nature as a produced thing. In brief, both society and nature are produced, hence malleable, transformable and transgressive". The most recent approaches in urban political ecology, a field of study that seeks to overcome the separation between city and nature (Heynen et al., 2006), regard urbanity as a process strictly related to suburban and rural spaces. These liminal spaces of relations between urban, natural and rural patches produce particular socionatures perfectly represented by protected peri-urban areas born of the socioecological relations that characterise the processes of urbanisation of nature, as are the cases illustrated here. In this chapter, we explore these areas in the context of Lombardy Region, focussing on the ways in which they contribute to the deconstruction of the nature–culture dichotomy, the dynamics of power and participation from which they derive and the processes of commodification or conviviality at place.

Lombardy: land of records

Lombardy is a land of economic, demographic, urban and environmental records. The region is a dense network of small and medium-sized towns, many of which are of considerable historical, artistic and cultural importance. Lombardy presents an articulated physical framework extending from the Apennines to the heart of the Alps, touching the upper and lower Po Valley and the foothills marked by the great pre-Alpine lakes: Garda, Iseo, Como and Maggiore.

Among the region's records is the system of protected areas, which covers almost 25% of its territory. The excellence of Lombardy's environmental policies can be attributed to five main factors: the timeliness of the initiation of legislative activity for environmental protection, the multifunctionality of the protected areas, their strategic geographical distribution, the network layout of the protected area system and the articulated typology of areas provided by regional legislation.

The region began intense legislation on parks and reserves as early as the early 1970s despite the lack of a national reference framework. Regional Law no. 58 of 17 December 1973 allowed the establishment of Ticino Park (1974), Nord Milano Park (1975), Groane Park (1976) and Colli di Bergamo Park (1977), often decades ahead of the initiatives undertaken in other regions of Italy.

The multifunctionality of the parks in Lombardy is another important aspect. Besides the classic objectives of natural heritage conservation, hydrogeological protection and the recreation and cultural development as well as scientific purposes, Lombardy's protected areas serve the purpose of "recovering natural and environmental assets", with pre-eminence given to "the development of agricultural, forestry and pastoral activities and other traditional activities that favour the economic, social and cultural growth of resident communities" (Regional Law 86 of 1983). This broad spectrum of purposes has been supported by special economic programmes. Among them, the Special Project for Agriculture, which allocated some €10 million between 2000 and 2011 to supporting agro-environmental measures in the protected areas, has played a strategic role.

Moreover, the geographical distribution of the protected areas has been far-sighted. In the first phase, it favoured areas of the greatest criticality and landscape/environmental degradation, such as urban margins and the extensive foothills conurbation, the most densely urbanised area of the Po Valley megalopolis (Turri, 2000). Today, the Milan conurbation is partly mitigated by a coherent system of parks: Ticino Park to the west, Agricultural Park to the south, and North Adda and South Adda Parks to the east. North of Milan, despite a lack of physical continuity between the protected areas, there are no fewer than eight small and medium-sized regional parks, which counterbalance the pre-eminent urban fabric of the Milan metropolitan area. Lombardy's regional parks also cover the large reservoirs of Alpine and pre-Alpine nature thanks to the establishment of numerous mountain parks since the 1980s. The primary corridors of ecological continuity represented by the great rivers (Ticino, Lambro, Adda, Serio, Oglio and Mincio), whose courses connect the protected mountainous areas to the foothills and lowlands, have also been placed under protection.

The protected areas' network layout is another reason for their excellence. For the most part, these areas are interconnected, forming a network that amplifies their potential for biodiversity conservation. This network design is more accomplished in the Milan metropolitan area, while it is less dense in eastern Lombardy, where environmental protection policies have encountered more difficulties.

Finally, the typology of protected areas articulated by Regional Law 83/86 has played an important role in Lombardy's nature protection system. The family of protected areas is divided, on a continuum of increasing territorial extension and complexity, into regional parks, nature parks, nature reserves, natural monuments, areas of particular natural and environmental importance and local parks of supra-municipal interest (Parchi Locali di Interesse Sovra-Comunale – PLIS). The latter are proposed at the "grassroots" level, that is by local administrations that, in agreement with neighbouring administrations, decide to protect a particular physical, morphological or landscape element (such as a hill, mountain, stream, agricultural landscape or wooded area) of their territories to guarantee its conservation, enhancement and use. Until the early 1990s, PLISs constituted a rarely used safeguard tool. Today, there are more than 100, which demonstrates the active sharing of policies on protected areas and the widespread need for environmental and landscape quality expressed by local communities and administrations. The diverse scales of the protected areas have enabled active environmental protection across scales: from the most complex environmental systems, such as mountain ranges and river shafts, to punctiform sites, such as waterfalls, caves, irregular boulders and simple monumental trees.

The 10th legislature of Lombardy Region closed with the approval of Regional Law 28 of 2016 (Reorganisation of the Lombardy's System of Management and Conservation of Regional Protected Areas and Other Forms of Protection in the Territory). The new legislative initiative aimed to economically and organisationally streamline the protected areas through the merging of previously separated institutions and/or closer management and administrative collaboration. However, these objectives were not achieved due to a lack of real economic support and differences in the definition of a territorial framework and merging methods between local authorities.

More recently, as part of the cultural itineraries promoted for Bergamo–Brescia Italian Capital of Culture 2023, Bergamo Hills Regional Park and Brescia Hills Park presented a rich cultural itinerary called "*I parchi verso BG/BS 2023*" (Parks for BG/BS 2023) that was aimed at involving the population and institutions in suburban parks as landscape laboratories, places of culture and high natural quality, tools for the well-being of citizens, social cohesion and quality of life, and ecosystems. The process involved more than 50 associations, universities, scientific institutions, administrations and protected areas and resulted in the signing of the Bergamo–Brescia Charter of Peri-Urban Protected Areas[1] (Casti & Ferlinghetti, 2024). The document, which is the result of this participatory process, highlights the primary role of parks in the redevelopment of urban peripheries, not so much for the general salvation of the planet but to make urban peripheries agents of a new habitability (Berque, 2022) of nature based on the principle of reciprocity to shift from a status of parasites/users to that of guests/symbionts of the planet. In practical terms, the charter intends to promote actions for territorial restoration, to support and strengthen the voice of citizens and associations with public administrations, to encourage agricultural development projects aimed at recovering disappearing knowledge and cultural practices and to regulate the semi-natural by enhancing it and making it an open-air laboratory of educational experiences. Although Lombardy is the top region in terms of soil consumption (14%; ISPRA),[2] it falls to 16th place if the number of inhabitants is taken into account and to last place if GDP is considered. Lombardy is also the first region in Italy to have enacted a law against land consumption (Regional Law n. 31/2014), which envisages a 20% reduction in agricultural land consumption compared to 2014 forecasts and a net transformation of 0% by 2025 according to EU targets.

Peri-urban protected areas in Lombardy

With a population of just under 10 million, Lombardy Region is characterised by a diffuse urban system. The urban fabric is concentrated between the Prealps and the plain, sloping down to the Po River. This system is the spatial result of a vital economic system that has made the region the most important economic area in Italy and one of the most prominent in Europe. Going beyond a first glance, the presence of the natural system inside this urban fabric emerges clearly and produces an interesting landscape in which urban, rural and natural areas are closely interconnected and which can be interpreted as socionatures (Swyngedouw, 1996). The entire region, from rural landscapes to the urban fabric, bears traces of the socionatural interactions linked to its economic development. As described in the previous section, the recognition of the naturalistic and ecological value of areas in and around the Lombard urban fabric occurred at a particularly early stage, perhaps in connection with the region's rapid economic development.

This relationship is also quite evident in the localisation of the protected natural areas of the region. To illustrate this, we conducted a GIS analysis to map the spatial interactions between protected natural areas and the urban fabric to determine where this socioecological relationship is most evident. The use of GIS software allowed us to select protected areas – natural parks, regional parks and local parks of intermunicipal interest (PLIS) – and to analyse the presence of urbanised areas inside and near the limits of these protected areas, creating a buffer of one kilometre and using regional spatial data on land use. The resulting image (Figure 12.1) illustrates the different degrees of urbanisation across the region's protected areas. The analysis reveals a concentration of the protected areas in the most urbanised areas around the cities

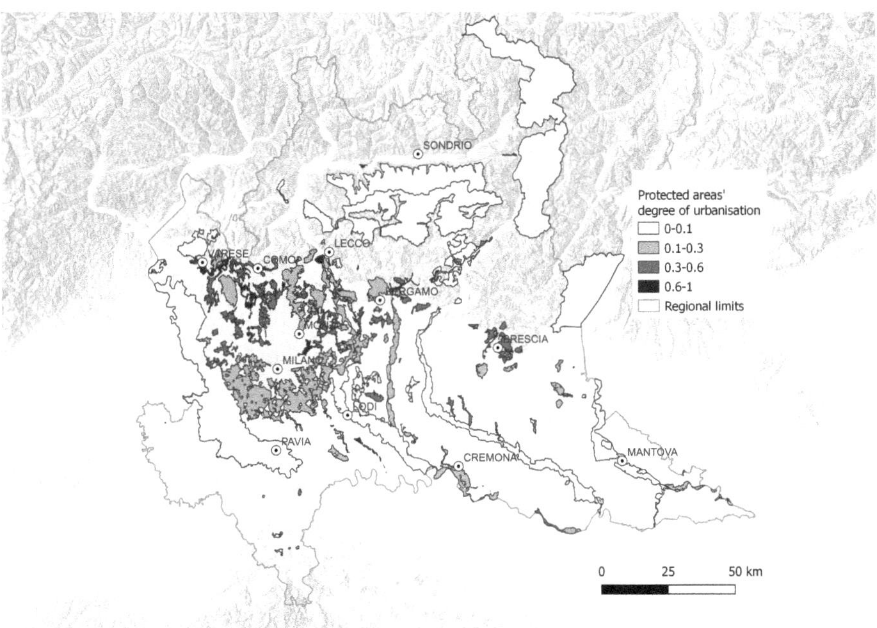

Figure 12.1 Degrees of urbanisation near and inside Lombardy's protected areas.

near the Prealps and along the rivers, which, among other factors, is motivated by the urge to contain urban expansion, as explained in the previous section.

Once we identified these areas, we enquired about the contemporary trends and challenges facing nature conservation and production therein. Specifically, we aimed to address three aspects. First, we were interested in exploring the relationship between the ecological and cultural values recognised in these protected peri-urban areas. In these contexts, perhaps more so than elsewhere, the interdependence of ecological and cultural values is quite evident. However, unlike in rural landscapes, in which this interdependence is more celebrated – for example, through references to biocultural heritage – in urban and peri-urban contexts, this interdependence is often hidden, contested or denied. Increasingly, urban nature is celebrated through calculations of ecosystem services, censuses of protected species and processes of de-paving or urban reforestation. This runs the risk of denying the domestication or coexistence processes at play, simply pushing human activities aside and reaffirming the dichotomy. This ignores the sociocultural processes on which such nature and its landscape dimensions depend. In other words, we wondered to what extent these areas contribute to the deconstruction of the nature–culture dichotomy.

Second, we considered it important to think in terms of power and resource distribution. On the one hand, the availability of public funds for conservation is constantly diminishing, a trend that leads to standardisation or the demand for more "efficient" management of protected areas. On the other hand, and as a consequence of this trend, there seems to be a search for greater participation of the private sector, associations and individual citizens in taking responsibility for certain aspects and practices of conservation in the form of increased civic stewardship.

The third issue concerns the progressive monetisation of peri-urban green spaces within economic mechanisms by offsetting carbon emissions and storing carbon or providing ecosystem services, not to mention their role in renewable energy production (e.g., photovoltaic parks). However, these processes coexist with experiences and places in which conviviality prevails over market dynamics. In fact, these areas are home to spaces and practices of conviviality that seek to operate outside the logic of the market, especially in relation to issues such as proximity, well-being and care, education and the production and consumption of organic and local food.

To examine how these processes take shape in the Lombardy context, we translated the above-mentioned questions into three concise research questions: What nature is being protected? What are the power dynamics involved? How is nature "valorised"? In the next section, we present two case studies conducted to answer these research questions. We selected these cases from the areas identified by the GIS analysis described above and on the basis of the prior knowledge and experience of the authors, who have long been involved in scientific collaboration and consultancy concerning these areas. We combined this insider's view, typical of an action research methodology, with more objective data, such as documents and regulations, information on the actors involved and completed or ongoing projects.

Bergamo and Brescia as case studies

Bergamo and Brescia are located in the densest agglomeration of the Po Valley megalopolis. Both are medium-sized foothill cities with a valley outlet and have grown thanks to the strong relationship between the mountains and the plain. Both are also experiencing a phase of post-industrial economic reconversion in which art, culture and nature play an increasingly important role in urban development strategies. A common characteristic of the two cities is the recognition of the ecological value of natural areas both inside and near the urban fabric (Ferlinghetti, 2013; Tononi, 2021). In both provinces, moreover, the tourism sector experienced strong growth, at least until the outbreak of the COVID-19 pandemic. As described in the following subsections, the two cases are also comparable in terms of nature conservation and production, as they have historically been surrounded by a rich countryside (in terms of cultural and religious sites and villages) that has contributed to the development of the cities themselves. Remnants of these *Corpi Santi*[3] (literally translated as "Holy Bodies", the name given to the areas outside the city walls) now constitute the core – sometimes neglected, sometimes celebrated – of the protected areas developed since the 1970s. In the next two subsections, we present the dynamics at play in these two contexts, structuring the description according to the three research questions.

Bergamo Hills Regional Park

Bergamo Hills Regional Park was created in 1977 with the aim of "safeguarding and improving the balance between nature and human presence in the hills surrounding the city".[4] It covers an area of more than 5,000 hectares spread over 11 municipalities. The area included in its perimeter has different physical and morphological characteristics, ranging from plains to hills and mountains, and includes both highly natural sites (Natura2000[5]) and contexts of great architectonic value, such as the mediaeval centre of Bergamo. Historically, the protection of nature in this area has centred on the cultural landscape and the protection of open and green spaces, woods, and private gardens or cultivated areas in hilly urban contexts. In fact, the land use plan of the 1950s was the first to enshrine a form of protection against land consumption, preventing owners from constructing oversized buildings on their plots. Between 1954 and 2022, land consumption was 9%, while in the surrounding areas, it reached the significant percentage of 90%.[6] In 2022, protection was extended to the lowland agricultural areas surrounding the city centre. The ecological value of this area is certainly recognised, as the park protects endangered habitats and species, but the prevalent narrative presents a "bucolic" rural landscape within the city as a kind of urban heterotopia.

This is therefore a context characterised by strong proximities and frictions, with protected habitats juxtaposed with transformative interventions for strong urban regeneration. Thus, there are many overlapping interests. In terms of governance, the park is managed by a board appointed by representatives of the municipalities involved and financed mainly by funds from the regional and provincial levels. Coordination and institutional cohesion appear to be a strength of the park, but this is offset by a lower level of civil society participation than in the case of the Brescia parks. Cooperatives are entrusted with agricultural land or the management of certain refreshment and environmental education areas, but power over the processes of protection, exploitation and valorisation firmly lies with various institutions, both public (the Municipality of Bergamo and its Botanical Garden and the University of Bergamo) and private (MIA Foundation and real estate companies). One of the results of this institutional cooperation is the landscape and environmental regeneration project Biodiversity in the City: Bergamo's Val d'Astino, promoted by the park and local public and private bodies, which won the seventh edition of the Council of Europe Landscape Prize.

How, then, is nature "valorised"? In short, one can say that the park follows at least three paths. First, nature enters valorisation processes related to the production and sale of "natural" food produced inside the park. Quality peri-urban agriculture and food biodiversity are key issues for the city, and we may say that this is an example of integrated socionature, in which the naturalness of the area is connected to the population through small but virtuous initiatives for local food production, trade and consumption. Second, recent years have witnessed the proliferation of naturalistic engineering interventions for climate change adaptation and renewable energy production initiatives (e.g., photovoltaic parks). In these processes, nature enters eco-technical enhancement processes, calculating ecosystem services and producing conflicts with those who support landscape conservation. The third process, which is more difficult to demonstrate but which we would like to mention as a possible risk, concerns green gentrification (Anguelovski et al., 2022; Ferlinghetti, 2013). Indeed, the enhancement of the urban nature of Bergamo risks increasing real estate prices and favouring processes of selection and expulsion of different inhabitants, practices and socionatures.

Brescia Quarries Park and Hills Park

Despite its industrial development, Brescia also appears to be a city with strong ecological sensitivity and several environmentalist movements and initiatives linked to the promotion of sustainability, especially as a reaction to the tragic effects of the presence of an industry such as Caffaro within the urban perimeter and excavation activities on the surrounding hills (Tononi, 2021). The process of establishing the Brescia Hills PLIS was initiated in 1996[7] and has recently been in an expansion phase, with the possible annexation of the park to Quarries Park and their proposed evolution into a regional park.

Data on land use and consumption provided by Lombardy Region[8] indicate that the two parks serve as areas of urban expansion containment. Between 1954 and 2021, the artificial surfaces of the city of Brescia increased from

18% to 48%, while urbanisation in Quarries Park and Hills Park stopped at 20% and 7%, respectively. The two parks are characterised by different dynamics of limiting urbanisation. In the former, an environmental regeneration of mining sites has been taking place quite recently. In the latter, there have been processes of conservation of natural areas. The two areas have in common a reduction of agricultural land, which in Quarries Park has been reduced from 95% to 48%.

Quarries Park is a PLIS established in 2018.[9] It covers 960 hectares and is characterised by the presence of quarry ponds resulting from mining activities, which have heavily modified the area. Here, too, there is a mix of different socioecological dynamics. The recovery and renaturalisation of the quarries intervene on a historical peri-urban rural landscape. The presence of ponds, wetlands, areas of spontaneous vegetation, buildings and canals – a sign of the pre-existing agricultural organisation – makes it a valuable area. Nevertheless, there is no shortage of reasons for criticism, such as the urban and industrial areas in the vicinity and the road and railway infrastructures across the territory. Regarding power dynamics, the park was created under pressure from local environmental associations, which for decades have demanded that the abandoned quarries be renaturalised and protected. Thanks to actions of the local community, a territorial context marked by intense environmental conflicts related to the presence of quarries and landfill projects has been transformed into a protected area of great socioecological importance for the entire city (Pietta & Tononi, 2021; Tononi, 2021). The associations' commitment has resulted in participatory management with the establishment of a management committee and a scientific committee in which the local community is represented and can interact with the director of the protected area chosen by the municipality.

Brescia Quarries Park is thus another example of a landscape in the process of becoming. From an agricultural area and then a mining site, it has now become a green area in a post-industrial context, usable and strongly experienced, despite conflicts between views on valorisation and protection models among the population. The park is not only a large area with ecological value but also a process of urbanisation of nature – a socionature that generates socioecological problems and conflicts (Tononi, 2021, p. 115).

Conclusions

In this chapter, we discussed the protected peri-urban areas of Lombardy as elements closely connected to the urbanisation of nature. The protection policies in the region have been particularly innovative in that they first targeted the most problematic peri-urban areas of the Po Valley megalopolis and then expanded to the "sanctuaries" of biodiversity represented by mountainous areas and river basins. Peri-urban parks have generated peculiar territorial dynamics. Particularly significant has been the containment of land consumption. However, these parks have also triggered widespread processes of urban regeneration and redevelopment that have generally been successful from a socioecological point of view, despite some criticisms. Exemplary in this respect are the cases of Bergamo Hills Park and Brescia Quarries Park and Hills Park.

The two case studies were motivated by similar needs: the protection of natural areas near and around the two cities, defined herein as socionatures produced by the urbanisation of nature during the years of economic and industrial development. These socionatures reveal the socioecological dynamics that create geographical evidence of the historical interactions between society and ecosystems. However, the two cities' paths to the protection and valorisation of these socionatures differ. Bergamo Hills Park recognises and protects peri-urban natures following a top-down process initiated by the main municipal institutions. This is a sociocultural process that, with the planning tools developed from the mid-20th century onwards, prevents urban speculation. Brescia Quarries Park and Hills Park, on the other hand, arose from a bottom-up intervention by civil society as compensation for the impacts of the industrial sector. Starting in the 1990s, local associations, through environmental conflicts and protests, requested the restoration and protection of some natural areas degraded, in the case of Quarries Park, by the human exploitation The result in both cities has been the recognition and valorisation of socionatures, which are evidence of the interactions between urban, rural and natural dynamics. Thus, the concept of urban socionature can be a powerful tool for unveiling the nexus between nature and urbanity. In this study, we employed it in connection with the current critical debate on nature conservation.

The recent Charter of Peri-Urban Protected Areas marks an advancement of strategies for the protected areas of Lombardy, moving from conservation and territorial regeneration to the promotion of a new relationship between local communities and environmental components to generate a new model of planetary habitability (Berque, 2022) centred on a renewed social and environmental centrality of urban margins. Such charters may draw renewed attention to the cultural and social role of "nature" in urban contexts, regarding protected areas not so much as sanctuaries of biodiversity but rather as laboratories for experimenting with forms of landscape co-construction.

Notes

1 https://www.comune.brescia.it/sites/default/files/2023-11/Firma%20Carta%20aree%20protette%20periurbane%20BGBS%20-%20La%20carta.pdf accessed 18 October 2024.

2 https://www.snpambiente.it/wp-content/uploads/2023/10/consumo_suolo_estratto_dati_2023_anni_2006-2022.xlsx accessed 26 November 2024.

3 In modern times, the designation *Corpi Santi* indicates the suburban area in some Lombard cities (Milan, Bergamo, Como, Vigevano, etc.). The toponym has medieval origins and probably indicated the areas where the first cemeteries with the burials of Christian martyrs were located. It later defined the rural properties of church and religious congregations, and subsequently it was used for a city's property outside the town walls.

4 https://www.parcocollibergamo.it/

5 https://eur-lex.europa.eu/EN/legal-content/glossary/natura-2000.html
6 DUSAF Lombardy Region data processed by R. Ferlinghetti and A. Azzini (CST, University of Bergamo).
7 https://www.comune.brescia.it/aree-tematiche/verde-e-parchi/parco-delle-colline
8 https://www.regione.lombardia.it/wps/portal/istituzionale/HP/DettaglioServizio/servizi-e-informazioni/Enti-e-Operatori/Territorio/sistema-informativo-territoriale-sit/uso-suolo-dusaf/uso-suolo-dusaf
9 https://www.comune.brescia.it/aree-tematiche/verde-e-parchi/parco-delle-cave

References

Anguelovski, I., Connolly, J. J. T., Cole, H., Garcia-Lamarca, M., Triguero-Mas, M., Baró, F., Martin, N., Conesa, D., Shokry, G., Del Pulgar, C. P., Ramos, L. A., Matheney, A., Gallez, E., Oscilowicz, E., Máñez, J. L., Sarzo, B., Beltrán, M. A., & Minaya, J. M. (2022). Green gentrification in European and North American cities. *Nature Communications*, 13, Article 3816. https://doi.org/10.1038/s41467-022-31572-1

Aschenbrand, E. (2024). How urbanization is shifting the context of nature experiences from economic to recreational. *People and Nature*. https://doi.org/10.1002/pan3.10594

Berque, A. (2022). *Entendre la terre: A l'écoute des milieux humains*. Le Pommier.

Büscher, B., & Fletcher, R. (2020). *The conservation revolution: Radical ideas for saving nature beyond the Anthropocene*. Verso Books.

Carr, E. (2000). Park, forest, and wilderness. *The George Wright Forum*, *17*(2), 16–30.

Casti, E., & Ferlinghetti, R. (Eds.), (2024), *La Carta di Bergamo-Brescia delle aree protette periurbane. Verso un'alleanza con la natura per l'abitabilità della Terra*. Grafo.

Castree, N., & Braun, B. (Eds.). (2001). *Social nature: Theory, practice and politics*. Blackwell.

Clément, G. (2004). *Manifeste du Tiers paysage*. Sujet/Objet. http://www.gillesclement.com/fichiers/_tierspaypublications_92045_manifeste_du_tiers_paysage.pdf

Cronon, W. (1996). The Trouble with wilderness: Or, getting back to the wrong nature. *Environmental History*, 1(1), 7–28. https://doi.org/10.2307/3985059

Ferlinghetti, R., (2011). Paesaggi minimi: Caratteri, valori, prospettive: Un approccio geografico. In M. C. Zerbi, & R. Ferlinghetti (Eds.), *Metamorfosi del paesaggio: Interpretazioni della geografia e dell'architettura* (pp. 103–116). Guerini.

Ferlinghetti, R. (2013). Spazi verdi ed evoluzione del paesaggio urbano a Bergamo: Una sintesi geografica. In M. A. Breda, & M. C. Zerbi (Eds.), *Rinverdiamo la città: Parchi, orti e giardini* (pp. 285–332). Giapparelli.

Heynen, N., Kaika, M., & Swyngedouw, E. (Eds.). (2006). *In the nature of cities: Urban political ecology and the politics of urban metabolism*. Routledge.

Loi, M., & Pierucci, G. (2025). Wild encounters: Toward a reinterpretation of urban green policies by the rivers in Rome and Cagliari. In S. Bonati, M. Cisani, M. Tononi, & G. Zanolin (Eds.), *Italian Natures. Perspective on the social nature in the "Bel Paese"*. Routledge.

Lorimer, J. (2015). *Wildlife in the Anthropocene: Conservation after nature*. University of Minnesota Press.

Metta, A., & Olivetti, M. L. (2019). *La città selvatica: Paesaggi urbani contemporanei*. Libria.

Müller, A., Bøcher, P. K., Fischer, C., & Svenning J.-C. (2018). "Wild" in the city context: Do relative wild areas offer opportunities for urban biodiversity? *Landscape and Urban Planning*, *170*, 256–265.

Piana, P., Brocada, L., Hearn, R., & Mangano, S. (2024). Urban rewilding: Human–wildlife relations in Genoa, NW Italy. *Cities*, *144*, Article 104660. https://doi.org/10.1016/j.cities.2023.104660

Pietta, A., & Tononi, M. (2021). Re-naturing the city: Linking urban political ecology and cultural ecosystem services. *Sustainability*, *13*(4), Article 4. https://doi.org/10.3390/su13041786

Scazzosi, L. (2020). Urban agriculture as heritage: Methodological issues and perspectives. In L. Scazzosi & P. Branduini (Eds.), *AgriCultura: Urban agriculture and the heritage potential of agrarian landscape* (pp. 17–44). Springer International Publishing.
Swyngedouw, E. (1996). The city as a hybrid: On nature, society and cyborg urbanization. *Capitalism Nature Socialism*, *7*(2), 65–80. https://doi.org/10.1080/10455759609358679
Tononi, M. (2021). Nature urbane: Rinaturalizzare la città (post)industriale, l'esempio di Brescia. *Rivista Geografica Italiana*, *2*, 102–118. https://doi.org/10.3280/rgioa2-2021oa12035
Turri, E. (2000). *La megalopolis padana*. Marsilio.

13 Wild encounters

Toward a reinterpretation of urban green policies by the rivers in Rome and Cagliari

Martina Loi and Ginevra Pierucci

Rereading urban green policy as beautification and control

The role of nature in urban policies – often focused on park-making projects in which "green" is presented as a self-evident concept and used as a catchy keyword – is increasingly crucial and is essential for improving quality of life and urban sustainability. While there is widespread acceptance of green urban policies as universally good and beneficial (Angelo, 2021), there is often a lack of critical thinking about their actual impact or what terms such as "nature", "green", or "park" actually mean.

According to Barua and Sinha (2022), urban nature can be divided into three groups: "cultivated", "feral", and "wild". The urban planning system tends to only allow space for productive and controlled "cultivated" nature, while "feral" nature, which remains a product of urban dynamics, and "wild" nature, which this chapter focuses on, are not welcomed in urban settings and are expelled through a superimposed green policy framework. This leads to the paradox in which an idea of "wild" nature can eventually be included in urban areas, but only within a controlled boundary and thus embedded within the same system. One such example is the *Biblioteca degli Alberi* in Milan, where the fields of "wild" flowers enjoyed by park users must be constantly managed to keep their wild appearance.

Urban nature is placed in a system of domestication that is focused on the control of human and non-human behaviors, in contrast to the (spatial and social) degradation (Angelo, 2019, 2021; Loughran, 2020) of wild nature or "uncivilized" people. Viewing both nature and parks as a moral good for urban improvement opens the door for green policies to be used as a control tool (Certomà, 2011) in which domesticated nature is associated with civilized and morally appropriate urban practices. Through this "green as good" direct association (Angelo, 2021), it is possible to implement exclusionary processes (excluding unwelcome human and non-human subjectivities) while also promoting land value extraction. It is possible to identify "grey epistemologies" (Finewood, 2016) behind green discourse, with green policies ultimately oriented toward maintaining the status quo and initiating new urbanizations.

DOI: 10.4324/9781003465164-17

Projects proposing the regeneration, protection, enhancement, or even "renaturalization" of urban nature can be seen as promoting the urbanization of nature or – moving from the classical definition of Urban Political Ecology (Heynen et al., 2006) – processes of urbanization *through* nature. Nature is framed and rendered aesthetically valuable through the superimposed view of the park as almost the only possible tool to provide nature to citizens. This approach supports subsequent real estate operations triggered by the presence of "good" nature in the city. As such, urban nature is appropriated by the urban planning system and delivered as a product by decision-makers.

While this system seeks to control urban nature, some forms of spontaneous nature still resist domestication. In this chapter, the concept of "wildness" as an epistemological and methodological device is used to construct a critically engaged geography of urban nature and discuss park-making processes. This framework is used to examine the undefined green spaces along two urban riverbanks: the Tiber river in Rome and the Is Cungiaus river in Quartucciu in the Cagliari metropolitan area. The approach used shifts the focus from these places as abandoned and degraded urban voids (as described by the policymakers) to show the tensions of these spaces with the park-making processes.

Urban wildness

The concept of "wildness" we use here for urban contexts has nothing in common with the current demands by some branches of ecology to reforest and rewild urban spaces, which seek to preserve pristine nature by recreating it. Rather, such concept originates in opposition to the idea of pure nature proper to "wilderness" (Birch, 1998; Cronon, 1996). Wildness extends its gaze beyond the boundaries of nature reserves and considers nature as a result of a co-production process by a variety of territorial actors including also humans, societies, cultures. It legitimizes the existence of a heterogeneous nature composed of complex assemblages of bio-social relations (Callicott & Nelson, 1998). In short, wildness profanes the pristine nature ideal associated with Romanticism, which founded the North American myth of the frontier and was carried through to the establishment of National Parks and the expulsion of Indigenous communities (e.g., Stan, 2014).

Some of the characteristics of "wildness" that differentiate it from "wilderness" include: spatial and social heterogeneity, which welcomes the contemporary connectivity of urban and rural dimensions; the fluidity of its configurations, being a mobile, always-changing subject composed of relations more than elements; and the autonomy of its wild inhabitants, which are intended to be territorial actors with agency in co-producing wildness (as well as the city itself) (Whatmore & Throne, 1998). This conception inspired Sarah Whatmore in *Hybrid Geographies* (2002) to refocus her study of wild nature – which is often associated with exotic places – to include all geographical dimensions, including urban contexts (Hinchliffe et al., 2005).

Wildness has become a key concept in Environmental Humanities, spreading beyond the North American context and encouraging a broader range of research, which isn't homogenous but changes according to the culture. In fact, the term "wildness" has different connotations in every language and does not even exist in many Indigenous languages (Klein, 1994). In the Italian context, wildness is represented as *selva*, a term etymologically related to "forest", "wild", and "savage". This has been explored by Environmental Humanities (Angelucci et al., 2021), mainly through architecture, and particularly in the city, where it has given rise to monstrous representations of urban landscapes (Metta & Olivetti, 2019; Metta, 2022; Carignani & Pierucci, 2022). The concept has also been investigated theoretically,[1] mainly through philosophy, in the new materialism framework inspired by Deleuze and Guattari. From this perspective wildness has been considered a relational device – as suggested by Vannini and Vannini (2019) – because it produces a mirror effect between what is represented as wild nature and what is considered to be a civilized society. In so doing, it reveals the cultural point of differentiation between humans and non-humans:

> Wild and savage could be considered as positional names, names of an infra and extra subjective relation, where they demonstrate what's at stake in the process of constitution of one's own perception, but also an infra and extra human relation, where they demonstrate how humans name those dynamics that constitute themselves in differing them from other humans and other non-human beings.
>
> (Gentili & Giardini, 2020, p. 95, translation by the author)

In the urban context, wildness gains even more radical connotations because it is increasingly abstracted from the dimension of "naturalness". Wildness here intersects the literature on interstices that redeem the discarded spaces of a city from consideration as urban voids, highlighting their socio-ecological richness and making them worthy of political agency (Sousa Matos, 2014). Interstices and wildness find common ground in the urban context, where the heterogeneity of nature's spatial configurations is fully expressed and where political resonance is given to the autonomy of its actors. Wild inhabitants emerge as outcast creatures that are socially rejected because considered producers of urban's decay. Focusing attention on such dynamics, using a wildness device, it emerges the political nature of the discourses sustaining park-making plans, which are often centered on the social degradation of urban nature (Barchetta, 2021) rather than on ecological restoration.

With this premise, this chapter investigates the outcast relations that exist between riverbanks' unplanned green areas and the city. Riverbanks (not yet made into parkland) represent the quintessential otherness in contrast to urban nature (Osti, 2021). Wildness relational device directs the observation focus on the spatial relations between the riverbanks and the city. This approach is enacted using a relational scale (Pierucci, 2023). The encounters taking place between the river and the city are considered as practices of daily dwelling, widening the spectrum of inhabiting and inhabitant subjectivities beyond the juridical definition of residence.

As a thought device, wildness directs our interpretation and shapes our common methodology, which is articulated through two complementary grounds. A critical reading of park-making plans illustrates the conception of space adopted by policymakers (e.g., degradation, emptiness, decay, etc.), together with the imagined and proposed future of the area. This analysis is combined with ethnographic fieldwork (Yi'En, 2014; Springgay & Truman, 2018), oriented toward observing the daily practices that constitute the relations taking place between riverbanks and the city. We enter these encounters, considering our research walks as dwelling practices more than explorations. The field descriptions in this chapter are the result of a series of crossings in and out of riverbanks. These encounters are a way to connect with, represent, and reinterpret the parallel city of urban nature (Phelps & Silva, 2018), navigate spatial and temporal frames, and link scales and locations (Sidaway, 2022). The two case studies are at different stages: one is still in the planning stage (Quartucciu), while the other has been completed (Rome). As such, it required the use of an additional research method: interviews were conducted during the Rome case study to understand the changes in perceptions of urban nature before and after the park was established.

Quartucciu: a fenced river, a park, and the transformation of practices

Quartucciu is a small town with an agropastoral history close to Sardinia's capital city of Cagliari and now forms part of the conurbation surrounding the Molentargius Pond, one of Europe's most important wetlands. Within this conurbation, Quartucciu's urban growth has been influenced by its compressed position between the larger towns of Selargius and Quartu Sant'Elena. The town is crossed by the Is Cungiaus river,[2] which flows into the Molentargius Pond and forms – together with two other rivers – part of Cagliari's hydrological network. This hydric system is very controlled and is in no way enhanced by metropolitan policies. The Is Cungiaus river is now enclosed for safety reasons, with high metal fences or guardrails along almost its entire urban length, and is partly covered over. Consequently, it is a rewilded river with no apparent urban function; placed in an interstitial position within Quartucciu's urban geography, its environmental and urban potential is completely denied, and it is instead considered a vector of pollution and degradation. Despite its fences, there are traces of usage of its riverbanks.

The river is currently the subject of a design competition launched and funded by the municipality. The competition is a "renaturalization" project to create a linear park with sports facilities, playgrounds, trails, relaxation areas, bars, and other amenities. The competition's document[3] not only highlights the project's requirements but also its critical points. The document starts from the assumption that "green" "should be desired as a widespread asset, and the [river] environmental corridor represents a strategic environmental axis for the redevelopment of degraded areas" (Comune di Quartucciu, 2021, p. 15). The riverbanks are expected to provide greenery, (spatial and probably also social) decorum, and connectivity between neighborhoods to "develop environmental protection and use it as a competitive advantage to promote new entrepreneurial initiatives" (Comune di Quartucciu, 2021, p. 18). Beneath the "renaturalization" and environmental protection surface, the river's nature is framed as a resource to trigger urban development.

The project is located in an area considered strategic for the reconstruction of Quartucciu's social and cultural identity and for enhancing its historical, cultural, and environmental heritage. To reconstruct this identity, the municipality is proposing a park that works as a control device promoting the domestication of nature but is detached from Quartucciu's agropastoral history. The project document recognizes that the river represents a boundary between the town and the productive countryside. However, while on the surface the project is honorably aimed at promoting environmental protection, new plans for middle-class housing are leveraging the future park to increase their attractiveness, demonstrating an overlap of private interests and public involvement. It is certainly true that historical territorial dynamics have been fractured by inconsistent urbanization and that the area is now a patchwork of built/unbuilt spaces and cultivated/uncultivated fields; however, interstitial traditional practices persist in this area without any acknowledgment.

Encountering Quartucciu's wildness

The project in Quartucciu is only at a preliminary design stage, and yet it is easy to anticipate that it will bring about a major change to the urban balance. The riverfront will become a new center of attraction, modifying urban values and providing a further push toward urbanization. This urbanization of nature risks completing the erasure of the historical territorial dynamics that still partly resist.

However, this project is currently only on paper. In this in-between state, it is possible to highlight issues with the project given the river's present condition. Through repeated encounters with the river's wildness over several months, it was possible to connect the project's premises to the urban conditions, practices, and possibilities (Loi, 2024) embedded in this interstitial space. While the fieldwork investigates the actual, the project analysis reveals a future envisioned by policymakers; however, these two viewpoints do not coincide, with the latter seemingly contradicting the former. At the same time, wild encounters can help rethink the very meaning given to wildness and provide potential tools to support project planning, particularly as the project is currently unrealized – as such, fieldwork observations can help point out the gap between the actual and the proposed.

The river is currently fenced off and virtually inaccessible; however, several riverside practices persist, highlighting an interesting parallel between the river's wildness and fluidity and the interstitiality of these practices. A series of chairs have been placed under several trees along the river, which are used by local people to meet while enjoying the shade. The chairs have been replaced over time, showing the repeated use of the riverside. It is also often possible to meet dogwalkers and elderly people strolling along the narrow grassy path outside the fence line – and inside where the fence has been torn away. These people seem familiar with the place and with each other because they use it to gather socially to talk about their daily lives and neighborhood news. Alongside these riverside activities are half-hidden economic activities, with fields still cultivated, others used for foraging or grazing sheep, small workshops, and informal businesses. Some of the empty areas have already been converted into small parklets with micro-furnishing operations, beginning the urbanization process.

Although it does not have an aesthetically pleasant nature by urban standards, these encounters with wildness show that these "empty" and degraded spaces "are not waiting to be filled with something but … are living spaces filled with meanings" (Sousa Matos, 2014, p. 65). There have been no attempts by the project designers to critically read the riverbank, its material and symbolic flows (Kaika, 2005), and its relationship with territorial dynamics. One of the declared project objectives is to regain the strong identity of the town and to mend its flaws; however, in light of these reflections, the proposed park risks simply achieving a "beautification" of the area that raises the land value and expels its former inhabitants. What is proposed as a "renaturalization" is indeed a process of urbanization of nature in the Urban Political Ecology sense. Rather than seeking a possible mediation between the planning requirements and territorial specificities, the park project superimposes a new spatial normativity that removes pre-existing uses without creating a new coherent and cohesive unity and instead only fosters further urbanizations.

Park-making policies along the Tiber river in Rome

Parkland is a core aim of the great political operation planned in the Tevere Master Plan for the Tiber River, which advocates for the reconnection of divided natural environments near the riverbanks by creating green corridors (Roma Capitale, 2023). The project anticipates 24 interventions – to be managed by the municipality and the region and mainly supported by Jubilee 2025 and National Recovery and Resilience Plan funds – most of which involve the construction of riverside parks.

This study focuses on one of these interventions, the Parco Tevere Marconi, built in the summer of 2021 and inaugurated in October 2022 and now open to citizens. The case study was carried out by analyzing the park project documents and entering the riverbanks, walking across and in them, and encountering its inhabitants. Field research and interviews were repeated before and after the park was completed to understand the differences in the city-river relations and in the perception of urban nature produced by the park.

Parco Tevere Marconi

The project is presented as a redevelopment of the environment and ecology of the area through remediation and extensive vegetation planting. However, these proposed actions are not supported through the inclusion of data on the ecological degradation of the area, while the reason given for the development is that the area has been "abandoned" and that is necessary to clean up the waste produced by "informal settlements". The underlying motivation for the park-making turns out to be more interested in the social than in the natural degradation.

> The work begins with the environmental and ecological redevelopment of the area by clearing it of the rubbish and waste materials accumulated by the squatter settlement, which will be completed by planting rich vegetation.
>
> (Translation by the authors, Regione Lazio, 2024)

Field observations during the park-making stage noted that the remediation operations included eradicating *Arundo donax* reeds and riverbank tree felling, with this existing vegetation substituted for new vegetation. In addition, the last act of the park-making stage was the construction of a fence – which had not been included in the project plan – enclosing the riverside park, with a gate to open the park only during daytime hours.

The park-making process placed the riverbanks under social and biological control in the name of a specific idea of nature considered suitable for the citizen's needs; however, no participatory process had been put in place to inform the project. As such, the project created an exclusive space and exacerbates the symbolic distinction between civilized urban nature and uncivilized wild nature.

Encounters by the Tiber

The Tiber riverbanks are divided from the rest of Rome. High walls called *muraglioni*, built between 1871 and 1926, protect the ancient city from floods (giving the riverbanks urban furnishings), while in the contemporary part of the city, the riverbanks are contained by embankments and fences whose goal is "to demarcate, as a matter of urgency, areas dangerous to public safety" (Translation by the authors, Roma Capitale, 2024). Added to these restrictions are other symbolic boundaries that contribute to this division, making the contemporary city's riverbanks almost invisible to its citizens, as was reflected in our interviews. After the park's inauguration, some of the park users interviewed seemed to have lost their memories of what had been there a few weeks before. Two of the eight people interviewed[4] stated that there had been "wasteland" there, while most of the interviewees stated that they had not noticed anything before the park was built – in one case, the participant affirmed directly that before the park, "there was nothing".

A relational approach, inspired by wildness, makes it easy to cross symbolic boundaries and find a path to follow (Pierucci, 2023). Suddenly, plants appear and devour barriers, sheep occupy the footbridge, fishermen cast their rods, and the rats of the Tiber invade the city at night. Multiple boundaries and accesses create various micro-geographies that give space intimacy to the heterogeneous mix of inhabitants and practices there. "In" and "out" references do not apply, as the space was neither public nor private: to inhabit has a wider meaning than to reside. There is no citizen status by the Tiber.

When the sun sets and darkness limits the view, smells and soundscapes reveal undeniable connections between the riverbanks and the city: the roaring of bonfires rises from the Tiber and fills the street with the smell of cooking while traffic soundscape harmonizes with the crickets to compose the silence of the river. From inside the riverbanks, the city is experienced through its scraps of surpluses and rejects. Remnants of synthetic lawn support the banks alongside bags, tires, and reed roots. Rushes exceed the urban metabolism and are used as building materials for informal settlements, animal burrows, artistic installations, and daily nourishment.

From this perspective, urban wildness represents the backstage for civilization, where everything that is expelled and disowned by the capitalist city cohabits: people, places, things, and ideas (Sassen, 2014). Its mirror image, the urban park, seems to work as an aesthetic mask hiding behind "green" landscapes, while the social and ecological struggles that are forced to move still exist nearby. Some of these people are Roma families; they have a great – but bad – reputation. The citizens' perception of the riverbank is that it is a "gipsy land" (as a park user told us during an interview), fueled by the institutional health services and the census data of the river population, which are exclusively focused on Roma. These people are currently facing a systemic eviction from the Tiber, justified in the media by degradation rhetoric and accusations of "involuntary" fires and by politically referencing the park-making of the Tevere Master Plan on the new riverbanks (RomaToday, 2024).

In truth, also Italian citizens and migrants inhabit the riverbanks in many different ways, such as Bruno and Cesare, two of the last eel fishermen, and Stefano, who recalls his childhood memories living by the river, and Rita, a retired landscape architect at the Municipality of Rome who loves to feed the river's wild animals. Interviewed after the park-making remediation, Rita exclaimed, "They have killed the Tiber!" when pointing to the tree-felling operations and the resulting flight of the animals. All of these people have left the now park-made riverbanks but still inhabit the river nearby. Their eviction and the loss of entanglements has had no media coverage, highlighting how these heterogeneous territorial actors and practices are invisibilized and muted.

Evidence from the field shows that park-making policies by the Tiber are creating social and biological control through the enclosure of the riverbanks and the expulsion of certain categories of beings and practices whose citizenship rights are not politically recognized.

Conclusions

In this chapter, we undertook a critical review of urban green policies and the potential exclusionary logic they pursue by analyzing two riverside parkland projects. At the Is Cungiaus river in Quartucciu, the river is fenced and public access denied, and the park-making process is oriented toward the construction of a renewed relationship between the river and its surroundings and the regeneration of the adjacent suburbs, which currently host a variety of informal practices. The second case study is the Tiber River in Rome, where an urban park has been placed on the riverbank with the declared purpose of redevelopment and ecological remediation but with the intended aim of contrasting with the supposed social degradation of the area. While they belong to different contexts, both projects have been pursued with the same logic of supplying good nature as a tool to control space and practices.

This park-making logic expresses the contradictions behind the management of urban nature. Urban green policy tends to deny certain riverside uses to "give back" riverbank spaces to citizens, managing these spaces in a way that frames and controls nature (e.g., by enclosing the areas of the park). Behind the aims of environmental protection and regeneration is the denial of other forms of dwelling, the expulsion of (marginalized) populations, or the eradication of undesired animals and plants for the benefit of the "civilized" population.

This reinterpretation of urban green policies has been possible thanks to the concept of "wildness" and the relational approach it inspires, which provided a basis for the "wild encounters" in these two contexts. Shifting away from wildness as a carrier of pollution and degradation, wildness becomes a privileged standpoint that helps reveal the logic of nature exploitation that green policy frameworks hide behind when voicing environmental protection aims. The mirror effect of wildness provides the means to encounter and observe practices that still resist urban appropriation, show socio-ecological richness, and build a parallel city in the less controlled interstices.

There is no space here for nostalgia toward such practices. Thanks to the Tiber case study, where the park has already been built, it is possible to show that park-making policies are not threatening the riverbank practices with extinction, as they still exist, but are simply masking, hiding, and denying them. What we question is not the project's aim but its use of a standardized urban nature management paradigm (the park) – in which nature is made accessible through a regulatory framework with defined spaces, permitted functions, and desired users, leaving nature anything but a pure element of urbanization – without a proper reading of the city as it is (or was) before the park was made.

Notes

1 The Italian national research project Sylva undertook important theoretical work on the concept. Some of the results are published in *Sylva* (https://mimesisbooks.com/index.php/mim/catalog/series/sylva; last accessed on 12.06.2024) and in a series of "*selva*" guides to several Italian cities published by Nero (forthcoming).

2 Is Cungiaus, meaning "the enclosed (farms)," is the name of the fenced fields that historically surrounded Quartucciu. The river takes its name from this territorial feature.
3 The "Documento preliminare alla progettazione" (Preliminary Project Document) sets out the project requirements for the design competition.
4 Eight of the people interviewed were park users: three females and five males, two of whom lived outside the area, three inside, and three nearby. Three people were in their 70s, one was in their 60s, one was in their 50s, one was in their 40s, and two were in their 20s.

References

Angelo, H. (2019). Added value? Denaturalizing the "good" of urban greening. *Geography Compass*, *13*(8). https://doi.org/10.1111/gec3.12459

Angelo, H. (2021). *How green became good: Urbanized nature and the making of cities and citizens*. University of Chicago Press.

Angelucci, D., Armiero, M., Giardini, F., Gentili, D., & Bussoni, I. (Eds.) (2021). *Environmental humanities*, Vol. 1. DeriveApprodi.

Barchetta, L. (2021). *La rivolta del verde. Nature e rovine a Torino*. Agenzia X.

Barua, M., & Sinha, A. (2022). Cultivated, feral, wild: The urban as an ecological formation. *Urban Geography*, *44*(10), 2206–2227. https://doi.org/10.1080/02723638.2022.2055924

Birch, T. H. (1998). The incarceration of wildness: Wilderness areas as prisons. In J. Callicott, & P. Nelson (Eds.), *The great new wilderness debate* (pp. 443–470). University of Georgia Press.

Callicott, J., & Nelson, M. (Eds.). (1998). *The great new wilderness debate*. University of Georgia Press.

Carignani, L., & Pierucci, G. (2022), Eco Bestiario Ligure. In A. Bertagna & M. Giberti (Eds.), *Selve in città* (pp. 164–169). Mimesis.

Certomà, C. (2011). Critical urban gardening as a post–environmentalist practice. *Local Environment*, *16*(10), 977–987. https://doi.org/10.1080/13549839.2011.592181

Comune di Quartucciu. (2021). *Lavori di rinaturalizzazione ambientale e idraulica e realizzazione del parco lineare sul Rio Is Cungiaus. Documento preliminare all'avvio della progettazione*. https://www.comune.quartucciu.ca.it/index.php/download/eyJpdiI6IldDZStjWlc0T1pxcjNHK0MwQXhJQkE9PSIsInZhbHVlIjoiOFQ1Q2p0OUJRYWN0VzFDcGFKbi9uOWgzcE52blVFazFFSHlwLzRvY25KQT0iLCJtYWMiOiJiZDdmMDEyOTBhNDFkZDM2MDY1ZWI1MDVmOGVkYTFmOTRlZTcwNDU0YTMwMjhhNzk4MDE4ZjVkNTU0MzhiMzlhIiwidGFnIjoiIn0=/01_dpp_rio_is_cungiaus_rev.pdf

Cronon, W. (1996). The trouble with wilderness: Or, getting back to the wrong nature. *Environmental History*, *1*(1), 7–28.

Finewood, M. H. (2016). Green infrastructure, grey epistemologies, and the urban political ecology of Pittsburgh's water governance: Pittsburgh's water governance. *Antipode*, *48*(4), 1000–1021. https://doi.org/10.1111/anti.12238

Gentili, D., & Giardini. F. (2020). Selva e stato di natura: variazioni cinestetiche per il contemporaneo. *Vesper*, *3*, 78–99.

Heynen, N., Kaika, M., & Swyngedouw, E. (Eds.). (2006). *In the nature of cities: Urban political ecology and the politics of urban metabolism*. Routledge.

Hinchliffe, S., Kearnes, M. B., Degen, M., & Whatmore, S. (2005). Urban wild things: A cosmopolitical experiment. *Environment and Planning D: Society and Space*, *23*(5), 643–658. https://doi.org/10.1068/d351t

Kaika, M. (2005). *City of flows: Modernity, nature, and the city*. Routledge.

Klein, D. (1994). Wilderness: A Western concept alien to Arctic cultures. *Arctic Institute of North America, 20*, 1–6.

Loi, M. (2024) Urban motorways as spaces of possibility: Urban interstices and everyday practices around a motorway in Sardinia. *Urban Studies, 61*(13), 2548–2564. https://doi.org/10.1177/00420980241240208

Loughran, K. (2020). Urban parks and urban problems: An historical perspective on green space development as a cultural fix. *Urban Studies, 57*(11), 2321–2338. https://doi.org/10.1177/0042098018763555

Metta, A. (2022). *Il paesaggio è un mostro: Città selvatiche e nature ibride*. DeriveApprodi.

Metta, A. & Olivetti, M. L. (2019), *La città selvatica. Paesaggi urbani contemporanei*. Libria.

Osti, G. (2021). *Fiumi e città. Un amore a distanza*. Padova University Press.

Phelps, N. & Silva, C. (2018). Mind the gaps! A research agenda for urban interstices. *Urban Studies, 55*(6), 1203–1222. https://doi.org/10.1177/0042098017732714

Pierucci, G. (2023). Foto–geografie nella selva urbana: studio sull'interazione tra selvatico e urbano presso il Tevere. In A. Pase, A. Bondesan, & S. Luchetta (Eds.), *XXXIII Congresso Geografico Italiano 'Geografie in movimento', 1 'Elementi, Animali, Piante. Mobilità dei costituenti, delle forze, degli organismi'* (pp. 224–227). Cleup.

Regione Lazio. (2024, February 2). https://progetti.regione.lazio.it/parcoteveremarconi/ Retrieved June, 6, 2024.

Roma Capitale. (2023). *Masterplan Tevere*. http://www.urbanistica.comune.roma.it/m–tevere.html. Retrieved June, 6, 2024.

Roma Capitale. (2024, February 13). *Transennamenti*. https://www.comune.roma.it/web/it/scheda-servizi.page?contentId=INF43709&pagina=3#:~:text=Il%20Municipio%20realizza%20transennamenti%2C%20a%20carattere%20provvisorio%2C%20finalizzati,di%20urgenza%2C%20zone%20pericolose%20per%20la%20pubblica%20incolumit%C3%A0. Retrieved June, 6, 2024.

Romatoday. (2024, June 28). *Blitz sul lungotevere, sgomberata la baraccopoli di San Paolo. Trovata maxi discarica*. https://www.romatoday.it/cronaca/sgombero–baraccopoli–lungotevere–san–paolo.html

Sassen, S. (2014). *Brutality and complexity in the global economy*. Harvard University Press.

Sidaway, J. (2022). Psychogeography: Walking through strategy, nature and narrative. *Progress in Human Geography, 46*(2), 549–74. https://doi.org/10.1177/03091325211017212

Sousa Matos, R. (2014). Urban landscape: Interstitial spaces. *Landscape Review, 13*(1). https://doi.org/10.34900/LR.V13I1.794

Springgay, S., & Truman, S. (2018). *Walking methodologies in a more–than–human world: WalkingLab*. Routledge.

Stan, S. (2014). *Indigenous people, national parks, and protected areas*. University of Arizona Press.

Vannini, P., & Vannini, A. S. (2019). Wildness vitality: A relational approach. *Environment and Planning E: Nature and Space, 2*(2), 252–273. https://doi.org/10.1177/2514848619834882

Whatmore, S. (2002). *Hybrid geographies: Natures, cultures, spaces*. SAGE.

Whatmore, S., & Throne, L. (1998). Reconfiguring the geographies of wildlife author. *Transactions of the Institute of British Geographers, 23*(4), 435–454. https://doi.org/10.1177/03091325211017212

Yi'En, C. (2014). Telling stories of the city: Walking ethnography, affective materialities, and mobile encounters. *Space and Culture, 17*(3), 211–223. https://doi.org/10.1177/1206331213499468

14 Performing animality in city branding

The case of the *Sartiglia* in Oristano (Sardinia)

Giampietro Mazza and Sara Bonati

Introduction

As the more-than-human literature states, the social construction of (urban) space goes through the interrelation between humans and non-humans, where both have an acting role (Whatmore, 2002). The comprehension of relational space links to, as outlined by Massey (2005), the mentioned relationship with the non-human. According to the animal turn in research, there is a need of viewing non-human animals as territorial actors, which contribute the shaping of the geographical space. In the second half of the last century, Carl Sauer (1969) emphasised how the domestication of animals strongly contributed to the development of cultural landscapes.

The social nature debate driven by Braun and Castree (2001) shows how geographical research must unravel the narratives of the relationship between nature and society as well as humans and non-humans, what kind of nature(s)/relationships these narratives produce, and what purposes, and socioecological effects create. This analysis starts from these evidences considering how inter-species power relationships build places (in this chapter especially cities) and bodies, and shape how we populate them.

A florid literature on such dynamics has been produced in animal and cultural geography, igniting a new interpretation of the relationship between human and non-human animals. However, studies on it need to be implemented in Italy, especially those concerning the shaping and meaning of urban spaces.

The research starts with the idea of animal's performativity (e.g., Geiger & Hovorka, 2015), considered as practices of exposing, staging, show-making, and consuming non-human bodies as forms of interspecies space production (Birke et al., 2013; Raber & Mattfeld, 2017). The aim is to discuss the role that non-human animals have had in shaping the city. Thus, this work sets a narrative between animal and urban geography, analysing the power relations between humans and non-humans, with the aim of highlighting their role in building urban imaginaries (Philo & Wilbert, 2000; Wolch, 2012) and local identities which contribute to build the city as a more-than-human space.

However, most of these processes are often only promoted with the aim of selling a specific image of the city. This can be ascribed to city branding

DOI: 10.4324/9781003465164-18

strategies, according to which performativity relates to the need for a city to sell an attractive image and be a product on the market (Vanolo, 2017). Nevertheless, in these practices, non-human animals become (and are confined to be) a resource to consume for human leisure, a way to use and reuse animal bodies.

Based on these premises, the case of the city of Oristano, in Sardinia, is investigated. The event of *Sartiglia*,[1] in particular, is used to analyse how animals' role is renegotiated in temporary sport events and how the related urban landscape transformations are used to create a different perception and imaginary of the city, which includes the relationship between human and non-human animals. Into this discourse, the event becomes also a moment of human-animal renegotiation of the meaning of the city. In particular, the role of horses is investigated to understand how they shape the way of inhabiting, constructing and representing the city of Oristano. To this scope, document and content analysis, participant observation and site observation are applied as research methods (Chan & Marafa, 2013).

Animals' performativity in city branding

Animals have always played a role in the development of city life. Their presence in cities has significantly impacted the configuration of spaces and the social dynamics of cities, giving rise to unique and fascinating performativity. In the past, for example, the horse was the main method of transport, and urban streets were marked by the sound of horses' hooves, contributing to the creation of a characteristic visual, sound, and smellscape.

Furthermore, the horse contributed to economic growth and urban development (especially in the medieval period), symbolising social status and power for the nobility. However, according to the critical animal study debate, animals' ability to inhabit the city has been frequently limited by humans' permission and domestication, favouring the idea of animals as "lively commodities" (Hubbard & Brooks, 2021; Rock, 2013), that favour(ed) processes of body consumption, utilitarian interpretation, physical and cultural confinement, and commodification (among others, Bear et al., 2017; Brooks, 2006).

Today, the way non-human animals are allowed to live in the city is related to their function which could be mainly led back to companion, leisure and consumption purposes.

Indeed, in modern times, the role of horses in cities has changed, but although they are no longer a widely used means of transport, their presence in urban settings continues to be mainly instrumental.

Horses are used in sporting and recreational activities, historical parades, and specific events, brightening urban spaces and attracting tourists and visitors. In other cases, they are employed in surveillance activities by the police, playing a role in maintaining public order. In general, in the European Union, official data count 6 million horses, most of them employed in sports (European Horse Network, 2012; Parvage et al., 2013).

Recent studies have highlighted the animal agency in urban space, pushing for a "trans–species" urban theory that focuses on how animals-humans relationships contribute to shaping cities (e.g., Hovorka, 2018; Hubbard & Brooks, 2021) and promotes a more inclusive perspective on urban planning (Maller, 2018).

This has questioned, among others, how domesticated animals have been forced to live in the city over time. Discussions have been raised, for example, on horse-drawn carriages and the relative welfare of animals used in urban activities and events.[2] These practices, analysed in the city contexts, help to read the socially built spatial borders between human and non-human animals, such as the daily practices that the different actants stage to re-negotiate their power relations.

Focusing on animals employed in urban sport events, which are the object of this research, at a global level, numerous traditional sports and cultural events involve the presence of animals. One of the most representative events is bullfighting, both in its original Spanish version and its subsequent export to Central and South America. According to the Italian Federation of Historical Games, numerous sport events involve animals in Italy, as well. In total, 15 *palio* events (competitions between the districts of a city) take place in the country, distributed in different regions from North to South. Nine of them involve animals, particularly horses. Most of these events occur in urban centres, which means to reorganise the urban space to ensure its usability for animals (Bindi, 2012).

The most popular event is the historical reenactment of the *Palio di Siena* (Mugnaini, 2013), which involves the participation of horses bred in Sardinia and Sardinian jockeys. Most of these practices have started several discussions regarding the ethical and cultural identification moment of the *Palio* tournament (Marvin, 2015), including the condemnation of violence against animals (Brandes, 2009).

Nowadays, sports events represent a "cost–effective manner" for promoting the urban image (Herstein & Berger, 2013) and perpetuating a utilitarian and consumeristic view on the city, as part of neoliberal policies of place development. According to Kavaratzis (2004), place branding mainly works on two levels: on the "outside" to secure visibility, and on the "inside" to reinforce local identity.

Brands are usually built recalling traditional and historical events. However, an increasing literature on critical city branding has shown the limits of this practice (among the others, Colomb & Kalandides, 2010; Vanolo, 2017). Colomb and Kalandides (2010) highlight how the process of re-imagining the city through city branding, can marginalise specific cultural aspects and characteristics, favouring a process of cultural commodification, or, as Kearns and Philo (1993) state, "manipulation of culture", where branding practices become a way to reiterate bourgeoisie control on the city.

Considering those introductory remarks, the following section aims to discuss how the process of city branding implemented in Oristano can be either as a vector to re-imagine the "authentic city", and as a reiteration for interspecies power hierarchies, shadowing non-human performativity out of the *Sartiglia* event.

Case study

Horses in Sardinia have played and continue to play a significant role in local culture and traditions, contributing to unique events happening on major religious celebrations and the *carnevale* period.[3]

The Sardinian carnival, for instance, is characterised by different events involving animals. The most characteristic and distinctive aspect of this festival is represented by animal masks, which change according to the micro area of the island of belonging. All those masks are associated with the agricultural and rural world, representing the persistence of ancient pagan celebrations (Alziator, 1956).

Sardinian carnival masks represent an image of the symbiosis between humans and animals; induced by a relationship of dependence that ties the shepherd to the same animals representing their economic activity (Atzori, 1986). As Atzori observes, the use of masks represents a process of anthropomorphism that makes humans more wild animals-like than domestic ones, requiring magical rituals to exorcise the negative human-animal relationship.

The event that best summarises the human-animal relationship in Sardinia is the *Sartiglia*, a ritual that takes place during the carnival celebrations in the city of Oristano. It is a spectacle of late medieval origin, a direct legacy of the Aragonese presence in Sardinia, realised in its early stages to celebrate the sovereign of the Crown of Spain, Charles V, and which still today represents a strong spirit of cultural and sporting community for the city.

The city centre becomes the "racecourse" where the horsemen and horses compete during the event. In essence, it is a ring race that takes place in the urban centre of Oristano on the last Sunday and Tuesday of carnival, during which the knights (*Su Componidori*), riding his/her horse, are called upon to pierce a metal star with a sword.

Furthermore, *Sartiglia* has contributed to building the idea of Oristano as the Carnival city in Sardinia, where horses are protagonists. This is part of a city branding campaign that aims to build the image of Oristano as a medieval city and the city of horses.

The *Sartiglia* reminds how the human/animal relationship has spanned the ages, representing a symbiotic and insightful connection between the two species. A bond that has been functional and cultural, with evident impacts in the arts, literature, travel, conflicts, events and, in general, in people's daily lives.

Analysing the *Sartiglia*, the masks summarise the transfiguration of the characters (knights) from human to non-human, also representing an almost mythological element that fits perfectly within the Sardinian carnival, which, in accordance with Turchi (2018), portrays the Dionysian cult strictly connected to the very meaning of the Sardinian term Carrasecare (*carnevale*), which literally means "tearing living flesh": an apotropaic rite that exorcises evil, just as the mask represents the birth of a new life (human and non-human).

Thus, the centrality of the binomial human-animal (horse) is strongly represented in Sardinia[4] not only by the *Sartiglia* itself but also by a series of events such as stakes and *pariglie* that accompany local festivities and that have influenced this connection over time, favouring the implementation of national rules on the use of equines during events and their protection (Atzori, 1998).

The role of *Sartiglia* in city performativity

The following thesis focuses on unfolding and explaining two major positions: how the *Sartiglia* contributes to Oristano's city branding and its performativity, and how *Sartiglia* is contributing to performing Oristano as an interspecies city and what are the limits of this narrative.

The method used for this research is a combination of qualitative methods (in particular document and content analysis), participant observation and site observation (Chan & Marafa, 2013). They are used to identify the symbolic construction of the image of the city (Vanolo, 2017) that has been based on different resources and especially online promotional materials and websites, public declarations given by the event promoters, social media sharing, official event documents, and customers' web reviews (Shields, 1991). Thus, two specific typologies of actors have been taken in consideration: local institutions and how they promote the city of Oristano through the *Sartiglia*, and participants to the event and how they perceive it. This has represented the selection criteria for information sources. About the second group, the analysis was based on data uploaded voluntarily on the web-review platforms (Zukin et al., 2017), in this case Tripadvisor platform, which, according to Spalding and Parrett (2019), represents the largest global travel and event website.

Oristano, the carnival city. The looking for authenticity

The *Sartiglia* takes us back in time (Bale, 1994) when sports and events were mainly practised in the streets and city centres. Most sports competitions experienced a peripheralisation in city life, moving them out of the city centre or creating closed spaces like stadiums to practice them. This did not happen with events like the *Sartiglia*. Indeed, the showcasing of the *Sartiglia* is characterised by (and live of) the number of fans and supporters who follow the carnival rite and the cultural and identity value that assumes for the local community. Until a few years ago, the spectators were effectively part of the performance, allowing them to participate to the race physically occupying the racetrack, and this contributed to create the idea of an open and participated performance. Today, maintaining the event in the city centre can be reconnected, among the others, to the purpose of pushing towards more and more tourist-based urban centres.

If, on the one hand, these events, like *Sartiglia*, are finalised to safeguard the "authentic" components of the material and immaterial heritage, on the other one, they become functional to the commodifying and touristifying of the city (Zukin, 1993). Thus, the urban centre becomes a spectacularised place, and the heritage represented by the event becomes the vector to promote the city's consumption by the tourist industry.

From this perspective, it becomes valuable to examine consumers' perception of the performance and to define how the *Sartiglia* becomes a tool for conveying the tourist and identity image of Oristano.

What emerges is the relevant presence of the event on the web and social media pages of both the City of Oristano and Sardinian Government. Thus, the selected data show that: on Instagram, as of 10/06/2024 there are over 33,700 images or videos linked to #*Sartiglia* (thousands more images and videos appear with #*Sartiglia* and the relative year of the event, #La*Sartiglia*, #Sa*Sartiglia*) and over 458,000 for Oristano, in which very often the carnival event is directly referred to. Although this data are mainly quantitative, it is useful to understand the tourist dimension of the *Sartiglia*.

As previously mentioned, it is an event that plays a significant cultural and tourist role for the city. In recent years, studies have proliferated considering the level of tourist satisfaction (Zhang et al., 2014) and highlighting how participants or attendees of the event define the image of both: the event and the city (Marine-Roig, 2017).

The binomial *Sartiglia*–Oristano is also evidence of the number of tourists' presences in the city. Although, tourism in Sardinia is a predominantly seasonal phenomenon, especially during the summer months, in Oristano in 2022, the month with the highest tourist coverage of accommodation facilities was February (94.86%, more than July 91.47% and August 93.03%),[5] the period in which the *Sartiglia* is held.

This was also confirmed in 2023, when more than 70,000 presences were recorded on the main days of the event, of which 30,000 on 19 February and 40,000 on Tuesday 21 February (against a total capacity of 20,000 spectators set up by the municipality of Oristano). In 2022, the annual arrivals in Oristano were 288,466 (850,517 presences[6]); this means that the *Sartiglia* attracts a high slice of local tourism in two days (about 8.2% of the annual total).

The results of the analysis of Tripadvisor data show that on the page on theatres and attractions in Sardinia, the *Sartiglia* ranks fourth, behind the island's three most important theatres.

When it comes to Sardinian regional level, the *Sartiglia* leads the board, placing itself at the first place, confirming the historical and cultural role played by the event and recognised by participants.

Thus, it is important to point out that in October 23, 2023, the *Commissione Consultiva Carnevali Storici* (Consultative Historical Carnival Commission) of the Italian Department of Culture funded the Oristano Foundation with 76,405 euros for the delivery of the *Sartiglia* 2023, placing it in first place among Italian carnivals with less than 500 years of history.

Finally, comparing the reviews of experiences in Oristano, again on Tripadvisor, the *Sartiglia* appears in first position, with 108 total reviews, of which 90.74% excellent, 9.26% very good, 1.85% average, only 0.92% unsatisfactory, and 1.85% terrible. Most of the reviews are provided by Sardinian customers (31.5%), followed by tourists from the rest of Italy (28.7%) and the province of Oristano (24.1). Only 3.7% are interstate tourists and for 12% it is not possible to determine their origin.

According to a qualitative analysis of the 108 reviews recorded in Tripadvisor done by local tourists, it appears that the *Sartiglia* is Oristano, and characterises its history and culture. 19 of the 108 reviews associate the event with the city, highlighting the community's sense of belonging, strongly fitting within Oristano's cultural and tourist narrative. It is often described as an event that needs to be experienced at least once in a lifetime and how, even in the eyes of tourists (5 reviews), a strong identification of the locals with the days of the event emerges.

Towards the interspecies city: Oristano as the horse culture city

In 2019, the councillor for local authorities and urban planning of the city of Oristano declared that Oristano had to "become the recognisable and authoritative capital of horse culture in Sardinia". The project has been defined as having a "characterising strength for the city of Oristano", and the institutions considered the plan as the way to solve the problems of a deprived territory. A way to go out of the "chaos" and promote development and tourism.[7]

Saying that, the horse assumes a crucial role, even more important than the rider's, becoming the ruler of liberation/redemption. In this representation, all the participants in the event, including horses, become part of the tourist product and actors in the staging of the "authentic city" (MacCannell, 1976), such as projections of city imaginaries.

This process, which differs, but is parallel in a way to (and feeds on) the other process that accompanies the *Sartiglia* (i.e., the construction of local identity) described above, produces a "commodification of bodies" that corresponds to the staging of bodies for consumption.

In this representation, bodies are used in a way that does not relate to their original function (Scheper-Hughes & Wacquant, 2003).

Looking at the reviews on Tripadvisor, horses emerge as "the real protagonists" of the event, whose role is essential to the value of the *Sartiglia*. Furthermore, the event website talks of "symbiosis and respect that is established between the horse and the rider during the different moments of the contest" as "one of the many sources of great charm offered"[8] by the *Sartiglia*.

Although, at first glance, this protagonism of horses during the *Sartiglia* could represent a "provoking contestation over access to public space" (Urbanik & Morgan, 2013) or, more likely, the representation of harmonious coexistence between horse and rider, in truth, how animals can inhabit Oristano city during the event appears more as an example of the human–focused perspective on urban development and the limits of non-human animals' accessibility to the contemporary city.

Both placing horses' performance at the centre and the human control of horses' access to the city and their performativity testify to this. In the *Sartiglia*, animals' performativity depends on humans' expectations to satisfy requests for efficiency, elegance, and civic order (Instone & Sweeney, 2014).

Animals must perform perfectly synchronously with the knight to ensure the success of the race and the performances. Thus, it is better to use "docile animals" that "know what to do".[9] The research on equilibrium and the figures that are created show, on the one hand, this strong relationship between humans and animals and the need for rules to be followed and obedience to ensure the execution.

This recalls what Bettany and Daly (2008) stated about dogs in cities when they coined the word "hu/dog" to mean the conjoined human–dog as the only permitted way dogs can inhabit the city. Similarly, horses in the *Sartiglia* are a conjoined reality with humans, especially with jockeys. Jockeys are masked during the event, and from the moment they become *Componidori*, they cannot put their foot on the ground, remaining on the horse until the event is finished, recalling the image of the centaurs.

This produces the construction of what Bale (1994) defines as a sportscape, which in the *Sartiglia* translates with the alteration of urban space (Pioletti, 2008) to ensure the accessibility of the city to the horses and to create a cojoined stage of the authentic city. Indeed, the *Sartiglia* requires that the urban landscape changes drastically, turning into a racetrack for the knights' challenges and a catwalk for traditional Sardinian costumes.

Thus, Oristano's main square, Piazza Roma, becomes the place for the dressing of *Su Componidori*, who wears a white dress and mask and the blessing of the horses and riders before the race.

The central Via Duomo becomes the racetrack for the star race, and grandstands seating areas are set up along the street for the public. Then, Via Mazzini, in Oristano's historical centre, becomes the stage for the *pariglie*, in which triplet riders perform spectacular figures. Further places are the *Sartiglia* stable, located in the parking area of Palazzo degli Scolopi in Piazza Eleonora, and the *Sartiglia* village, which occupies a large area of the historic centre, including piazza Cova, via Diego Contini, and piazza Eleonora. The streets of the city centre and their different uses become the sacred site of the event. Thus, the city becomes a spectacle as a ground for authentic experience (Tiwari, 2010).

Furthermore, the streets are covered by fine, white sand, which is laid to make the horse race safest and more spectacular. Then, the sand is removed after the end of the event in a cleaning operation involving the *Sartiglia* Foundation and the municipality. The sand will then be moved back to the quarries, in the area from where it was previously sourced.

In the 2023 edition, has been decided to use two layers of sand: underneath the classic white stratum, apologetic sand will be added, especially at some critical points of the course, to ensure greater safety for horses and riders. This followed several accidents, like in 2016 when a horse died due to the road pavement and its slipperiness during a rainy day.

Although these actions were justified by a better protection for visitors and horses, they remind us that the city is not a place for animals. In this practice, the idea of the inaccessible city newly appears.

Into the imaginary building of Oristano as an interspieces city, animal's welfare was raised as a central point, especially in relation to the *Sartiglia*.

Indeed, a new statute of the event has been produced with this purpose in 2009. Accidents and deaths of horses during sporting and cultural events are considerably high and have moved public opinion and governance to adopt measures and regulations to eliminate the phenomenon and ensure respect for the rights of both riders and horses.

Further challenges for animals have been observed before and during the competition. An agreement between the *Sa Sartiglia* Foundation and the Department of Veterinary Medicine of the University of Sassari was signed in 2012. It created the bases and a guidelines document that provide for the constant monitoring of animals in all the phases of the event to ensure their welfare (Figures 14.1 and 14.2).

Conclusion

This research analysed the case of the *Sartiglia*, discussing the role of animals employed in urban sport races in city performativity and how they contribute to the cultural and tourist image of cities.

The example has contributed to understanding the place of "performative" animals in the city (Instone & Sweeney, 2014). The human–horse connection and the performativity of the horse in urban space represent a fundamental dimension of urban life in many localities. Although the role and function of the horse in the urban context have changed throughout history, the bond between humans and horses remains intact, testifying to the persistence of a millennial connection that transcends the barriers of time and space.

Thus, the chapter has discussed how the interrelationships between human and non-human animals contribute to the success of the city's "mise en scene", where the *Sartiglia* represents a moment of encounter and (con)fusion. During the event, the non-human world (or the part of this that is authorised) obtains the scene, becoming an "urban performing actor". This has consequences on the processes of urban (re)territorialisation, transforming the city into a place for horses, according to the multi-territorial dynamics described by Haesbaert (2012). Indeed, during the event, the city changes parts of its places' functions according to the race, such as its internal dynamics.

However, what emerges is the idea of the "showcase city", and especially of its historical centre, as more and more a place for leisure and tourism. In these dynamics, the identity function of the event becomes itself part of the show.

These relationships are evident analysing participants' perceptions of the event found on the Tripadvisor platform. According to the reviews, the city is represented as a place of celebration and tradition in which the entire community is involved, capable of engaging and fascinating the participants. The *Sartiglia* becomes much more than an event; the performing symbol of Oristano's cultural identity.

In this performance, the way animals inhabit the city is bound by rules from human society, making animals mainly an attraction in the city branding process. Thus, the city continues to be narrated according to an anthropocentric

Figure 14.1 The itineraries of the *Sartiglia*.

Authors' elaboration with google earth pro.

Figure 14.2 Image of the *Sartiglia*.

Source: Archivio Fondazione Oristano.

perspective that represents animals mainly "as Figures that so often remain backgrounded, stereotyped, or homogenised within both popular discourse and social research" (Van Patter, 2022, p. 2217). Into this spectacle, the risk is that horses are confined to be socially and culturally constructed bodies that perform in cities and contribute to the city performativity.

To conclude, this chapter tried to prove how a more-than-human perspective can help in better understanding lights and shadows in city branding (Vanolo, 2017). However, some questions would require further investigation, especially what happens to non-human animals when the lights go out and what remains of the interspecies city.

Notes

1 As will be seen more clearly below, La Sartiglia is a traditional equestrian joust held in Oristano (Sardinia) during Carnival.
2 In the Palio di Siena, Italy's most famous equestrian event, 43 horses have died as of 2023. The most recent was in 2018.
3 "Carnevale" is a festive period when people dress up and enjoy themselves before Lent, celebrating the reversal of social rules.
4 Among the most significant examples are the Ardie di Sedillo (OR) and Pozzomaggiore (SS) and several Palios, such as the Palio dell'Assunta in Orgosolo (NU), the Palio de sa Itria in Gavoi (NU) and the Palio dei Comuni in Fonni (NU).
5 Data Sardegna Turismo: http://osservatorio.sardegnaturismo.it/it/dashboard/dati–2020

6 Data Sardegna Turismo: http://chrome-extension://efaidnbmnnnibpcajpcglclefindmkaj/http://osservatorio.sardegnaturismo.it/sites/default/files/2024-06/OsservatorioTurismo_Report2023_25052024.pdf
7 https://www.comune.oristano.it/it/novita/news/Oristano-capitale-del-cavallo-in-Sardegna-un-villaggio-per-i-cavalieri-00001
8 Description provided by the event website: https://www.lasartiglia.it/it/articles/204/i-cavalli-della-sartiglia-di-oristano.html
9 https://www.corriere.it/bello-italia/notizie/sartiglia-oristano-sfida-cavallo-fatta-pazienza-scaramanzia-a17e1336-a23c-11ed-9cb0-cff1df2b4279.shtml?refresh_ce

References

Alziator, F. (1956). Presupposti allo studio del carnevale sardo. *Lares*, *22*, 49–55.

Atzori, M. (1986). Uomini-animali nelle tradizioni popolari della Sardegna. In Museo Nazionale… (Eds.), *L'uomo selvatico in Italia* (pp. 187–198). Museo Nazionale della Arti e Tradizioni Popolari.

Atzori, M. (1998). *Cavalli e feste. Tradizioni equestri della Sardegna.* Asfodelo.

Bale, J. R. (1994). *Landscape of modern sport.* Leicester University Press.

Bear, C., Wilkinson, K., & Holloway, L. (2017). Visualizing human-animal-technology relations: Field notes, still photography, and digital video on the robotic dairy farm. *Society & Animals*, *25*(3), 225–256. https://doi.org/10.1163/15685306-12341405

Bettany, S., & Daly, R. (2008). Figuring companion-species consumption: A multi-site ethnography of the post-canine Afghan hound. *Journal of Business Research*, *61*(5), 408–418. https://doi.org/10.1016/j.jbusres.2006.08.010

Bindi, L. (2012). A cavallo della tradizione: Palii, giostre e processioni lungo la penisola. *Voci*, *IX*, 27–41.

Birke, L., Bryld, M., & Lykke, N. (2013). Animal performances: An exploration of intersections between feminist science studies and studies of human/animal relationships. In M. Wyer, M. Barbercheck, D. Cookmeyer, H. Ozturk, & M. Wayne (Eds.), *Women, Science, and Technology* (pp. 495–506). Routledge.

Brandes, S. (2009). Torophiles and torophobes: The politics of bulls and bullfights in contemporary Spain. *Anthropological Quarterly*, *82*, 779–794. https://doi.org/10.1353/anq.0.0076

Braun, B., & Castree, N. (2001). *Social nature: Theory, practice, politics.* Blackwell.

Brooks, S. (2006). Human discourses, animal geographies: Imagining umfolozi's white rhinos. *Current Writing: Text and Reception in Southern Africa*, *18*(1), 6–27. https://doi.org/10.1080/1013929X.2006.9678230

Chan, C. S., & Marafa, L. M. (2013). A review of place branding methodologies in the new millennium. *Plance Branding and Public Diplomacy*, *9*(4), 236–253. https://doi.org/10.1057/pb.2013.17

Colomb, C., & Kalandides, A. (2010). The 'be Berlin'campaign: Old wine in new bottles or innovative form of participatory place branding?. In G. Ashworth (Ed.), *Towards effective place brand management: Branding European cities and regions* (pp. 173–190). Edward Elgar.

European Horse Network. (2012). *The European horse industry in the European regions key figures 2010, 2012.* http://www.europeanhorsenetwork.eu/index.php?page=horse-industry-in-europe

Geiger, M., & Hovorka, A. J. (2015). Animal performativity: Exploring the lives of donkeys in Botswana. *Environment and Planning D: Society and Space*, *33*(6), 1098–1117. https://doi.org/10.1177/0263775815604922

Haesbaert, R. (2012). *O mito da desterritorialização do "fim dos territorios" à multiterritorialidade.* Bertrand Brasil.

Herstein, R., & Berger, R. (2013). Much more than sports: Sports events as stimuli for city re-branding. *Journal of Business Strategy*, *34*(2), 38–44. https://doi.org/10.1108/02756661311310440

Hovorka, A. J. (2018). Animal geographies II: Hybridizing. *Progress in Human Geography 42*(3), 453–462. https://doi.org/10.1177/0309132517699924

Hubbard, P., & Brooks, A. (2021). Animals and urban gentrification: Displacement and injustice in the trans-species city. *Progress in Human Geography*, *45*(6), 1490–1511.

Instone, L., & Sweeney, J. (2014). The trouble with dogs: 'Animaling' public space in the Australian city. *Continuum*, *28*(6), 774–786. https://doi.org/10.1080/10304312.2014.966404

Kavaratzis, M. (2004) 'From city marketing to city branding: Towards a theoretical framework for developing city brands.' *Place Branding*, *1*(1): 58–73.

Kearns, G., & Philo, C. (Eds.). (1993). *Selling places: The city as cultural capital, past and present*. Pergamon Press.

MacCannell, D. (1976). *The tourist: A new theory of the leisure class*. Schoken Books.

Maller, C. (2018). *Healthy urban environments: More–than–human theories*. Routledge.

Marine-Roig, E. (2017). Measuring destination image through travel reviews in search engines. *Sustainability*, *9*(8), 1–18. https://doi.org/10.3390/su9081425

Marvin, G. (2015). The art of fierceness: The performance of the Spanish fighting bull. In L. Orozco, & J. Parker-Starbuck (Eds.), *Performing animality: Animals in performance practices* (pp. 39–56). Palgrave Macmillan.

Massey, D. (2005) *For space*. The Open University, Sage Publications.

Mugnaini, F. (2013). Le feste neo–medievali e le rievocazioni storiche contemporanee tra storia, tradizione e patrimonio. *Lares*, *79*(2–3), 131–158.

Parvage, M. M., Ulén, B., & Kirchmann, H. (2013). Corrigendum to "A survey of soil phosphorus (P) and nitrogen (N) in Swedish horse paddocks". *Agric Ecosyst Environ*, *178*, 1–9. https://doi.org/10.1016/j.agee.2014.08.017

Philo, C., & Wilbert, C. (Eds.) (2000). *Animal spaces, beastly places: New geographies of human–animal relations*. Routledge.

Pioletti, A. M. (Eds.). (2008). *Luoghi tempi e numeri dello sport. Un approccio multidisciplinare a un fenomeno complesso*. Patron editore.

Raber, K., & Mattfeld, M. (Eds.). (2017). *Performing animals: History, agency, theater*. Penn State Press.

Rock, M. (2013). Pet bylaws and posthumanist health promotion: A case study of urban policy. *Critical Public Health*, *23*(2), 201–212. https://doi.org/10.1080/09581596.2012.749343

Sauer, C. O. (1969). *Seeds, spades, hearths, and herds: The domestication of animals and foodstuffs*. MIT Press.

Scheper-Hughes, N., & Wacquant, L. (2003). *Commodifying bodies*. Sage.

Shields, R. (1991). *Places on the Margin. Alternative Geographies of Modernity*. Routledge.

Spalding, M., & Parrett, C. L. (2019). Global patterns in mangrove recreation and tourism. *Marine Policy*, *110*, 1–8. https://doi.org/10.1016/j.marpol.2019.103540

Tiwari, R. (2010). *Space–body–ritual: Performativity in the city*. Lexington Books.

Turchi, D. (2018). *I carnevali e le maschere tradizionali della Sardegna*. Newton Compton Editori.

Urbanik, J., & Morgan, M. (2013). A tale of tails: The place of dog parks in the urban imaginary. *Geoforum*, *44*, 292–302. https://doi.org/10.1016/j.geoforum.2012.08.001

Van Patter, L. E. (2022). Individual animal geographies for the more–than–human city: Storying synanthropy and cynanthropy with urban coyotes. *Environment and Planning E: Nature and Space*, *5*(4), 2216–2239. https://doi.org/10.1177/25148486211049441

Vanolo, A. (2017). *City branding: The ghostly politics of representation in globalising cities*. Taylor & Francis.

Whatmore, S. (2002). *Hybrid geographies: Natures cultures spaces*. Sage.

Wolch, J. (2012). Anima Urbis. In L. Harrison (Eds.), *Architectural theories of the environment* (pp. 227–248). Routledge.
Zhang, H., Fu, X., Cai, L., & Lu, L. (2014). Destination image and tourist loyalty: A meta–analysis. *Tourism Management*, *40*, 213–223. https://doi.org/10.1016/j.tourman.2013.06.006
Zukin, S. (1993) *Landscapes of power. From Detroit to Disney World*. University of California Press.
Zukin, S., Lindeman, S., & Hurson, L. (2017). The omnivore's neighborhood? Online restaurant reviews, race, and gentrification. *Journal of Consumer Culture*, *17*(3), 459–479. https://doi.org/10.1177/1469540515611203

Part IV

Geo-literary representations of Italian natures

15 Il Bel Paese by Antonio Stoppani

A geoliterary approach

Davide Papotti

Introduction

Italy, from a political point of view, is a rather "young" country. While the nation has been perceived as traditional due to its long and illustrious history and as a place "full of history" (a label that can still be found in tourist guides), its existence as a unified country is relatively new (from 1861). This gap between a consolidated cultural identity (based mostly on the Italian language and its literary and artistic tradition) and its recent political status has been a central issue in the national social debate since the immediate aftermath of the unification, which took place in 1861 and was territorially (and symbolically) completed with the military seize of Rome in 1870. Renowned politician and writer Massimo D'Azeglio once said, "We have made Italy. Now we must make Italians",[1] but from a cultural geography perspective,[2] it could also have been stated, "We have made Italy. Now we must make Italian landscapes". Even better, the order of the noun and the adjective could be switched: "Now we must make landscapes Italian". The landscapes of the nation, depicted and glorified in so many literary and artistic traditions, had to be "re-interpreted" and "re-read" as cultural constituents of the newly born political identity.[3] The "readability" of Italian landscapes played a role in the construction of the national identity.

To answer this research question, we focus on a book called *Il Bel Paese* (translated literally as *The Beautiful Country* or, more poetically, as *The Land of Beauty*), published by Italian scientist and priest Antonio Stoppani in 1876, a few years after the unification of Italy, when the process of consolidating both the national school system and the publishing industry was at its peak. After providing the basic information necessary to understand the historical and cultural context in which Stoppani's book found enormous success, I explain how *Il Bel Paese* contributed to the processes of consolidation of the Italian conscience of nature, sensibility to landscape, and territorial foundational narratives. Finally, I briefly show how *Il Bel Paese* became a widely used brand with different meanings in various contexts.

DOI: 10.4324/9781003465164-20

The historical context of *Il Bel Paese*

Il Bel Paese was one of the most successful long-selling books in the early history of unified Italy.[4] The author (Lecco, 1824–Milan, 1891) was a priest, geologist, palaeontologist, university professor, director of the Civic Museum of Natural History of Milan, president of the Italian Society of Natural Sciences, president of the Italian Geological Society, and scientific communicator.[5] He was the author of several scientific publications (especially about the region of Lombardy). *Il Bel Paese* was conceived as a work suitable for a larger audience and for school students. It is structured as a series of lectures (29 in the first edition, 34 in the 1881 edition) that an uncle gives to his nephews, illustrating the geography and the geology of the newly unified country, together with the regional differences. Together with another work that was much less successful in the following years (Cesare Cantù's *Attention. Reflections of a Commoner*), *Il Bel Paese* won the *Reading Book for the Italian People* national prize in 1877.

The syntagm chosen for the title of this book associates the word "paese" (which historically indicates "a large stretch of territory, mostly cultivated or inhabited, characterized by specific physical, climatic, economic or human features"),[6] with the adjective "*bel*" (beautiful), thus suggesting the primacy in aesthetic quality of the territory of the Italian peninsula. The title chosen by Stoppani establishes a clearly recognizable link with the most illustrious Italian literary heritage. The expression can be found in two of the three "crowns" of medieval Italian literature. Dante uses the expression in his *Divine Comedy*: "del bel paese là dove il sì suona" (*Inferno*, XXXIII, 80) to indicate the linguistic area where Italian is used (Italy would only be unified as a political entity more than five centuries later).[7] Petrarch uses the expression "bel paese" in his *Canzoniere*, following another nuance[8] in the meaning of the word: "il bel paese ch'Appennin parte e 'l mar circonda et l'Alpe'" (CXLVI, vv. 13–14). Petrarch sketches a synthetic overview of the physical "skeleton" of the peninsula, first identifying the internal division of the country due to the presence of the Apennines mountain range,[9] and then recalling the land and maritime borders of the country. In Petrarch's use of the expression, if compared to Dante's, there is a clearer definition of the territorial extension of the "bel paese," which opens the reader's imagery to the geographical profile of the future political entity. Stoppani quotes Petrarch's passage directly in his *Preface* without explaining whose work it is, a clear sign of trust that readers can identify it without further explanation. Petrarch's quote was put in the frontispiece of many editions, next to the title.[10]

It is important to recall the historical and cultural origins of the expression "il bel paese" to fully understand the cultural, social, and political resonance of the title of Stoppani's book. The book itself can be considered both a scientific textbook and a narrative text.

Stoppani's preface to *Il Bel Paese* (entitled *To the Preceptors*) clearly identifies the book as a text conceived and constructed for a teaching environment: "The scientific and philological notes that enrich the text were made necessary by the idea that this text could possibly be accepted as a textbook in primary and teaching schools".[11] At the same time, its literary quality is evident. Although the author limits his stylistic ambitions to a humble statement – "As for the style, I will try to be clear, and as for the language, correct" (*Preface*, p. 20) – the volume generally presents refined stylistic levels in terms of sophisticated lexicon, variety of adjectives, elegant syntactical constructions, and frequent use of similes and metaphors. Redondi (2014, p. 123) stated that "it was the writing style, strongly literary and even lyric, that made this *Bel Paese* of his [Stoppani's] a book of popular science that was completely different from the others".

For the didactic purpose of the book, the prose was intended to be simultaneously scientifically informed, communicative, and stylistically charming. The result is a text of a "hybrid" nature that can be approached with the conceptual tools offered by the most recent research in literary geography. Recent contributions by prominent scholars in the field, such as Sheila Hones (2024), propose to blur the traditional duality that considers scientific texts and literary ones separately. Stoppani's book, from this perspective, is a perfect example of a text that can conjugate the two dimensions, thus offering a fruitful field of inquiry for literary geographies.[12] For this reason, using the literary text as a "diagnostic index", as Tuan (1976) suggested, it is possible to understand which characteristics the "idea of nature" had in the scientific debate of the second half of the 19th Century and the role that natural heritage could play in the consolidation of the Italian national identity.

Il Bel Paese as a social construction of Italian natural heritage

The author starts the manuscript from the reassertion of a traditional classic motto, γνῶθι σεαυτόν, dating back to ancient Greek culture (it was sculpted on the pediment of the temple dedicated to Apollo in the town of Delphi) but used here by Stoppani in the Latin version (already a form of re–appropriation and "Italianization" of the Socratic saying) of "Nosce te ipsum". The idea that the newly born Italy should learn to "know itself" permeates Stoppani's literary project, as if the text could work as a "mirror", allowing the relatively young national entity to admire its beauties.

In the *Preface*, the author of *Il Bel Paese* openly claims that natural sciences must concur in "providing nutrition to popular literature" and "completing the knowledge that a population must possess of itself". Stoppani claims that in the Italian market in his time, it was easier to find books about the geography of foreign countries than works dedicated to the national territory. The same could be said of Italian scientists, who are often more knowledgeable about distant lands and do not know much about their homeland. Against this xenophilia, Stoppani states that it is necessary to devote attention to the Italian territory. The bases of *Il Bel Paese* thus go deep into the terrain of patriotism, of a rhetorical construction of a national identity that must be reinforced through the study of the country's physical geography.

The progression of Stoppani's statements about the nature and goals of the book is worth analyzing in detail. The following is a key passage (from the *Preface*) that frames the conceptual and rhetorical context of the book:

> By narrating what he saw and heard, the author of the book is conscious of having guaranteed what modern literature requires: the *truth*. By reducing it to a series of family conversations that he uses to share with relatives and friends, he could also be flattered by the fact that the book does not lack the virtue of *naturalness*. As for the *entertainment* and the *teaching*, this means putting together what is *useful* with what is *sweet*, which must be sought in popular literature. The author thought that he could be successful if he remained within the field of his specialty studies. Furthermore, if he desired that which is pasture of the intellect to become nourishment for the heart, and had a motivation toward moral enrichment, he had to ensure that man would never be separated from nature, or nature from man.
> (p. 20; italics in the original)

First, the author refers to an ancient and consolidated literary *topos*, the need for a scientific discourse to be "true". The statement that the author narrates "what he saw and heard" can be found in works by classic authors such as Herodotus. The closeness to the truth must reassure the reader that the narration is trustworthy. Specifically, this passage can offer an interesting perspective on the meaning that the world "nature" assumes in Stoppani's reasoning. The virtue that the author relates to the choice of the narrative frame (the idea of presenting evening chats between the uncle and his nephews) is presented as a means to assure "naturalness". The metaphorical use of the term "nature" is of particular interest because it serves as an introduction to the occurrence of the term that follows, which is inscribed in an ethical dimension.[13] After a further statement about the didactic value of the book (which can conjugate "what is useful with what is sweet"), Stoppani states that, to be sure about the effectiveness of his work, he had to remain in his realm of knowledge – that is, natural sciences. To describe the strict relationship between the enrichment of the intellect and the communication of emotional content that could also become "nourishment for the heart", the author uses an environmental metaphor, speaking of a "pasture" for the intellect. The geographical imagery that is thus activated serves as the perfect transition to the subsequent statement, which advocates the inseparability between "nature and man". The construction of the new national identity, according to Stoppani, was based on the education of the new generation in natural sciences, which can explain the complex dialogue between the natural environment and human activities.

It must be remembered, clearly, that Stoppani's conception of nature was theologically based.[14] He saw nature as an expression of divine action: "Yes, nature is the most universal and intelligible expression of God's essence" (*Preface*). The above-mentioned inseparability between "nature and man" comes from the fact that both are divine creations.

Il Bel Paese as a cultural construction of a national Italian narrative

Stoppani's choice of title for his work emphasizes the primacy of Italy in the realm of beauty. Showing an accurate knowledge of the current literary offers, the author mentions works by Jules Verne, the famous French author who wrote a number of bestsellers based on geographical imagery, such as *Journey to the Center of the Earth* (1864), *Twenty–Thousand Leagues under the Sea* (1869–1870), and *Around the World in Eighty Days* (1872). Mentioning the contemporary debate about literary genres and reflecting on the relationship between science and literature, Stoppani distinguishes two forms of novel: the historical novel (which in Italy is exemplified by *The Betrothed* by Alessandro Manzoni, published in 1827) and the scientific novel, which in Italy, "still has to find its Manzoni" (*Preface*). From a comparative literature perspective, Stoppani praises the richness of scientific production about the geography of the neighbouring country of Switzerland. While this country boasted a rich and lively production of scientific prose, it could not compete with Italy in the variety of its landscapes. The cultural construction of Stoppani's narrative discourse about Italy is based on the variety of Italian mountain peaks, seashores, volcanoes, islands, glaciers, and rivers, and its different kinds of soils, morphologies, climates, and vegetations. Italy is praised as a "geographical summary" of different natural environments, as a small epitome of the world.

The subtitle of *Il Bel Paese* is *Conversazioni sulle bellezze naturali, la geologia e la geografia fisica d'Italia*. The extended explanation (common at the time) in the subtitle confirms the goals that the author had in mind while writing the book: to conjugate the aesthetic appreciation of natural beauty with the scientific knowledge that allows us to understand it in its complexity. Once again, the presence of the adjective "natural" is central: from Stoppani's perspective, the natural environment could be fully understood only with the gaze of an informed human who studies and analyzes it.

The public utility of a widespread education about nature includes not only cultural aspects but also economic ones, as suggested by Visconti:

> Convinced that naturalistic knowledge of one's own country could be useful in the search for raw materials for the economic and social development of a modern nation, he [Antonio Stoppani] tried to demonstrate how it could be possible to approach such knowledge through observation, data collection, and descriptions of the large and small natural manifestations that are available to everyone and that can give everyone dignity and joy.
>
> (Visconti, 2012, p. 135)

Il Bel Paese as a brand

The linguistic formula *Il Bel Paese*, borrowed by Stoppani from the highest Italian literary tradition, was perfectly suitable for becoming a brand. If today we ask an average Italian what *Il Bel Paese* is, one of the most common answers will surely be "a kind of cheese".[15] In 1906, in fact, the founder of the Italian cheese factory Galbani (a commercial brand that still exists today) decided to produce a national cheese that could compete on the market with the then-dominant (in the upper section of the commercial offer) French cheeses. According to the entrepreneur, "It was not without a real disappointment that in the past, while waiting in front of the window of the main delicatessen shop in our towns, one could not see any luxury cheese with an Italian name. I was the first one who, after a long experience, managed to overcome the foreign import, commercializing my luxury cheeses, following the French tradition".[16] To emphasize the Italian nature of this new product, Egidio Galbani decided to name the cheese after the famous and popular book published a few decades earlier by Antonio Stoppani, whose portrait appeared on the packaging, together with a symbol that could evoke the national entity: a map of Italy.[17] The "*Bel Paese*" literary formula had already been transformed into a commercial brand.[18]

Concluding remarks

Today, Stoppani's bestseller is no longer widely read. Even though recent editions of the book[19] allow us to easily find the title on bookstore shelves, the popularity of the book is of course incomparable to its popularity at the end of the 19th century.

However, the expression *Il Bel Paese* remains widely known. It can be easily found, for instance, in tourism marketing, where it is frequently used as an appealing synonym for Italy. It is also prevalent in artistic works, such as the photographic series by Stefano De Luigi that became an exhibition for Palazzo Ducale in Genoa in 2023 and a book for the publishing house L'Artiere (De Luigi, 2023).

From a different perspective, the expression *Il Bel Paese* is sometimes provocatively used to speak about the decline of the Italian nation. In this ironic sense, the expression was used in cinema in 1977, when director Luciano Salce directed a comic (and at times grotesque) movie, entitled "*Il...Belpaese*" about the protagonist's (played by well-known Italian actor Paolo Villaggio) troubled life conditions in Milan during the 1970s (the so-called "years of lead" of terrorism).

Exemplary, from this perspective, is a newspaper article by historian and journalist Ernesto Galli della Loggia, significantly entitled "*Il Belpaese è diventato brutto*" ("The Belpaese became ugly"). In it, the author denounces the collapse of the main educational institutions in contemporary Italy.

We cannot know if Antonio Stoppani would agree with the pessimistic statement by Galli della Loggia if he had the opportunity to travel in today's Italy. Nevertheless, we must recognize the contribution he made to the construction of a national identity. In providing interpretative keys for understanding the richness of Italian natural environments, Stoppani aimed to contribute to the creation of a sense of belonging to the newly unified country.

Through his innovative work, which achieved both scientific accuracy and literary *charme*, he attempted to provide a solid "territory" to the rhetoric of the Italian identity. If we read *Il Belpaese* from our contemporary perspective, we can also find new resonance in his invitation to practice what we would today define as "tourism of proximity": the idea that, instead of travelling to the most remote areas of the world, we should learn to appreciate the proximal beauties of our national landscapes. Stoppani's description of the admirable diversity of Italian natural environments offers a scientific version of the tension between centrifugal forces of plurality and centripetal drives toward unity that still permeate the national political arena. His invitation to study and learn to appreciate this richness, by providing scientific foundations for aesthetic appreciation, remains valuable today.

Notes

1 The expression can be found in D'Azeglio's autobiography, *I Miei Ricordi* (*My memoirs*), published posthumously in 1867 (*Killinger*, 2002).

2 For an interesting overview of the role of landscapes in the consolidation of national identities in Europe in the nineteenth and twentieth century, see Renes, 2022.

3 Stoppani's work must be framed within a group of books that had the same goal: to build a national conscience of the territory. As Baffi (2002, p. 280) stated,

> *Il bel Paese* est sans doute le premier d'une longue série de livres qui, prenant prétexte d'un ou plusieurs voyages et dans des registres fort différents depuis le scientifique jusqu'au merveilleux, seront en réalité des ouvrages didactiques destinés à faire connaître la géographie, l'histoire, les traditions d'un pays.

4 Baffi (2002, p. 280) reminds us of the considerable success of *Il Bel Paese*, which has been a best seller for decades:

> D'autres éditeurs, après Agnelli, publièrent l'ouvrage: Cogliati, SEI, Barion, Hoepli, Vallardi, Trevisini, Salani… Le nombre et la rapidité des réimpressions et rééditions sont impressionnants surtout à partir de la fin du siècle: 1889: 7e; 1897: 46e; 1901: 50e; 1906: 58e; 1908: 77e; 1913: 86e; 1917: 110e; 1948: 150e. Les éditions économiques étaient particulièrement fréquentes, mais on remarque aussi des éditions dans lesquelles le texte est accompagné d'annotations de géographes et de reproductions photographiques, comme celle de 1908.

5 For an overview of Stoppani's biography and of his contribution to the scientific debate, see Zanoni, 2014.

6 The definition is taken from the Treccani dictionary https://www.treccani.it/vocabolario. All the English translations from sources originally in Italian, including Stoppani's texts, are mine.

7 For more information about Stoppani's interest in and knowledge of Dante's work, see Millefiorini, 2021.

8 "In a narrower sense, it is an extension of territory formed within specific borders and structures as a political organism; in this meaning, it is a synonym of State" (Treccani dictionary https://www.treccani.it/vocabolario).

9 The idea that in the past, before the unification brought the Southern question to the forefront, the main perceived division of the Italian territory structured along the longitudinal axis would be worth an in-depth analysis. It suffices here to quote the following passage by Marco Belpoliti, writer, journalist, and professor of Italian literature at the University of Bergamo:

> In fact the real geographical separation in the peninsula is not the north/south one, but the east/west one. The boot [a consolidated metaphor to indicate the Italian territory, due to its form on a map] is much more sloped than it appears in many of the maps hanging from the walls of classrooms, and one can go from north to south on the Adriatic coast without crossing any mountain range: from Venice to Bari, there is no clear separating border, except the river Po, which is not a real dividing line between north and south (Emilia and Romagna are southern regions). Geographers have long explained that the real geographical difference in Italy is the one determined by the two sides, the Adriatic Sea and the Tyrrhenian Sea, even though it is evident that the cultures follow geographical altitudes, and the differences between the villages located on the eastern slope and the western one in the Apennines are never sharp.
>
> (Belpoliti 2006)

10 As it happens, for instance, in many editions of the publishing house Cogliati.

11 For quotations taken from *Il Bel Paese*, I will not indicate the page number, but only the chapter, to make the reference easily traceable in every edition. I used as a reference text the 1915 edition by Cogliati (Milan; 94th paperback edition).

12 Another perspective that shows the newness of Stoppani's work is the use of the term "Anthropozoic," which has been interpreted as a precursor of the concept of "Anthropocene" (Luciano & Zanoni, 2023).

13 The ethical dimension of geographical knowledge is a recurrent theme. In his *Preface*, Stoppani also describes his chapters as "physical and moral sketches".

14 For more information about the religious goals of Stoppani's work, see Millefiorini (2015, pp. 50–54).

15 With regard to the fact that few Italians today remember the origin of the expression "*Il Bel Paese*," Baffi (2002, p. 278) states,

> Tous les Italiens connaissent l'expression *Il bel Paese*, devenue synonyme d'Italie, mais ils sont peu nombreux à savoir aujourd'hui que sa consécration vient du succès d'un ouvrage dont c'était le titre et qui a connu une grande diffusion entre la fin du XIXe et le milieu du XXe siècle.

16 Quote taken from Galbani's website: https://www.galbani.it/prodotti/bel-paese/storia.

17 The cheese was also registered with the name "Italico," emphasizing again the desire to achieve a national identity (and market).

18 Baffi (2002, p. 282) reminds us that the recent acquisition by a French corporation of the brand Galbani resulted in the disappearance of the portrait of Antonio Stoppani:

> Mais en 1993 la société Galbani, rachetée en 1989 par le groupe français Danone, décidait de procéder à une opération de « restyling »: les détails qui apparaissaient sur la carte d'Italie, en particulier la configuration orographique et le réseau routier, disparaissaient, l'appellation « formaggio bel paese » devenait tout simplement « bel Paese » et surtout le portrait d'Antonio Stoppani était supprimé. C'était le signe que la figure de Stoppani n'était plus reconnaissable, comme elle l'avait été pendant des décennies, et surtout qu'elle ne représentait plus un « argument de vente » et ne permettait plus d'identifier le produit.

19 See for instance the edition published by Theoria (a publishing house base in Sant'Arcangelo di Romagna, in the province of Rimini) in 2018, and the last one, published as recently as March 2024, by the Turin based well-known publishing house Einaudi (edited by Walter Barberis).

References

Baffi, S. (2002).'Fare gli Italiani': *Il bel paese* d'Antonio Stoppani. *Italies*, *6*, 277–299.
Belpoliti, M. (2006). A muro duro. *La Stampa*, 20 June, 2006.
De Luigi, S. (2023). *Il Bel Paese*. L'Artiere.
Hones, S. (2024). *Interspatiality. Inhabiting literary geography*. The University of Chicago Press.
Killinger, C. L. (2002). *The history of Italy*. Greenwood Press.
Luciano, E., & Zanoni, E. (2023). Antonio Stoppani's 'Anthropozoic' and the concept of Anthropocene. *The British Journal for the History of Science*, *56*, 103–114. https://doi.org/10.1017/S0007087422000590
Millefiorini, F. (2015). Il dialogo tra letteratura e altri saperi nel 'Bel Paese' di Antonio Stoppani. *Rivista di letteratura italiana*, *33*(3), 35–58.
Millefiorini, F. (2021). Antonio Stoppani lettore di Dante. *Studi sul Settecento e l'Ottocento. Rivista Internazionale di Italianistica*, *16*, 97–108.
Redondi, P. (2014). Le ascendenze svizzere di un libro diventato un best–seller: Il bel Paese di Antonio Stoppani. *Il Cantonetto*, *61*(3–4), 123–132.
Renes, H. (2022). *Landscape, heritage and national identity in modern Europe*. Springer.
Tuan, Y.-F. (1976). Literature, experience and environmental knowing. In R. G. Golledge, & G. T. Moore (Eds.), *Environmental knowing* (pp. 260–272). Dowden & Ross.
Visconti, A. (2012). Conoscenza e bellezza della natura nelle immagini del Bel Paese. In P. Redondi (Ed.), *Un best–seller per l'Italia unita. "Il bel Paese" di Antonio Stoppani con Documenti Annessi* (pp. 135–170). Guerini.
Zanoni, E. (2014). *Scienza, Patria e Religione: Antonio Stoppani e la Cultura Italiana dell'Ottocento*. Franco Angeli.

16 "*Mare crespo, vento fresco*"

Evaluation of the seascape in *I Malavoglia* by Giovanni Verga

Enrico Squarcina

Introduction

As Predrag Matvejević states

> We do not discover the sea by ourselves, nor do we view it exclusively through our own eyes. We also see it as others have seen it, in the images and tales that they have left us: we come to know it and recognise it at the same time.
>
> (1993, p. 143)

This statement highlights the role of popular culture in the evaluation and attribution of values to spaces, and marine spaces in particular. While human geography studies the relationship between human beings and physical spaces, the role of narration in influencing this relationship cannot be overlooked.

Starting from the writings of John Kirtland Wright (1947), academic geography focused increasing reflection on the role of literature in making sense of the world, a research field that became particularly lively following geography's humanistic turning point in the '70s. Moreover, that same period saw an ever-increasing number of literary scholars begin to turn their attention to the role played by geography in shaping the literary narrative.

Marc Brosseau (2022) identified a series of geographical approaches to literary texts: the first to emerge consists in the use of literary text as a descriptive complement to regional geography; the second, linked primarily to humanistic geography, concerns the use of literature to bring experience of locus and thus sense of place to the fore; the third approach, linked more to critical geography, uses literature to shine a light on society's relationship with space and the resulting inequalities; the fourth proposes a reflection on the evolution over time of the relationships between literature and geographical culture.

In addition to these approaches, we must also consider reflection on literature's ability to shape personal and collective sense of place and territory; indeed if, as Giuseppe Dematteis claims, geography does not describe the territory, but produces it (1985, p. 106), literature too, whether highbrow or popular, plays a part in how we see, interpret and subsequently interact with spaces.

DOI: 10.4324/9781003465164-21

In the literary field, starting from the '90s, interest in the representation of spaces involved the "study of the relationship between literature and the physical environment" (Glotfelty in dell'Agnese, 2021) which was defined with the term "Ecocriticism". This type of approach seeks to analyse the meanings attributed to the environment and to nature by individual authors or in entire literary genres, and aims to evaluate how the concepts of environmental discourse are used in literature[1] (dell'Agnese, 2021). Combining the ecocritical approach with the knowledge – postulated by the post-structuralist theoretical framework – that every form of writing plays an active role, insofar as it has the ability to change the world, to give an interpretation of it that can either validate or refute the prevailing ideology, produces a research approach that Elena dell'Agnese calls ecocritical geography (dell'Agnese, 2021).

These methods of analysing the relationship between geography and spaces are not mutually exclusive, which allows the construction of a dialogue that involves not only narratives and geographers, but also the ways in which the latter approach narratives.

In this work, I will attempt to make an ecocritical analysis of an Italian literary classic, *I Malavoglia* (*The House by the Medlar Tree*), written by Giovanni Verga and published in 1881. In particular, I will set out to show how the text in question describes the sea and how this physical environment constitutes far than a mere backdrop to the narrated events: indeed, on the one hand it acts as a metaphor for the life and social interactions of a 19th-century Sicilian community, while on the other it is a vehicle for the author's political and social ideas. Creating a parallel between a natural element which, as such, is considered indisputable, and the political and social structure of a community, allows the latter to be cast as acceptable, beyond criticism, natural.

The historical–ideological context Verga's ideology

Given how famous the novel in question is, we can dispense with a plot summary, as it will suffice to recall how the story revolves around the misfortunes of a fishing family from Aci Trezza, in the province of Catania, comprised of the head of the family, Padron 'Ntoni, his son Bastianazzo, who is married to Maruzza, referred to as la Longa, and their children: 'Ntoni, Luca, Mena, nicknamed sant'Agata, Alessi and Lia. The story takes place in the early 1860s and the events are triggered by the family's ill-fated attempt to climb the social ladder. Verga actually intended the book to form part of a projected five-novel series which would recount as many doomed endeavours to achieve social betterment, a series tellingly named *Ciclo dei vinti* (*The Vanquished*), but the author only completed the first two novels.

Verga's work reflects his pessimism on the one hand, and on the other his political conservatism which led him to be:

> highly critical of society not [...] in the name of a vision predicated on scientific grounds or a progressive idea [...] it is based above all on a constant harking back to the moral values of the past [...] In Verga's oeuvre

> the suffering and struggles of the poor are never denounced as a product of social injustice, of a situation that must somehow be changed: on the contrary, pain and hardship are the conditions that ennoble the existence of the poor. Such lives are "exemplary" precisely because they are tempered by suffering and need: it would therefore be absurd and counter-productive to seek to change the social and economic framework within which they play out.
>
> (Carnazzi, 1978, pp. 52–59)

Thus, the author's sincere commiseration with the condition of his humble characters is not an expedient employed to encourage subversion of the prevailing social order, motivated by progressive thinking; instead, it aims to hold up the moral virtues which, in his view, are being eradicated by the burgeoning bourgeois way of life.

The book in question describes the life of a village which, despite being very insular, suffers the first consequences of advancing modernity, and where an albeit modest social stratification pits the most wretched, namely those who struggle to survive, against those who lead a relatively privileged life and are driven by utilitarian self-interest. The author identifies this latter characteristic with the modern world, whose forward march is sustained by a gradual detachment from natural dimensions because "those at the bottom of the pyramid are still close to the natural dimensions of existence" (Carnazzi, 1978, p. 77).

The main characters of the novel, all members of the family nicknamed *Malavoglia* (the Reluctant Ones), are fisher-folk, meaning they live in close contact with the natural element that is the least tameable of all, the least amenable to territorialisation (Turco, 2010): the sea. Verga never offered any explanation as to why he chose this particular job for his protagonists, but perhaps it is precisely because of its close ties with this natural element.

Despite having struggled for social betterment through a business venture that ultimately leaves the family destitute, the head of the family, Padron 'Ntoni, never relinquishes his close bond with the natural dimension. This, in the author's view, is at the root of his stubborn determination to keep the family unit together, and of the honesty and loyalty that in their small world, taking its first tentative steps towards modernisation, are no longer virtues but, as noted by Vittorio Spinazzola (1977), factors destined to plunge the family into ruin.[2]

The narrative technique

The narrative technique employed in the novel warrants some consideration because, as Marc Brosseau states, expression of thought and geographical knowledge is by no means detached from the form of the text, given that language is more than the dress of thought, and geographers must be reminded of the importance of giving serious consideration to form (2022, p. 84).

The novel is considered the finest example of literary realism, a movement that set out, in the latter half of the 19th century, also following the rise of

French naturalism, to describe reality as objectively as possible, while most importantly concealing the voice of the author.

To achieve this aim, Verga adopted a technique described by Giacomo Devoto as impressionist (1954), using a range of narrative styles: the events are narrated in turn from the point of view of each of the various characters, but also by a choral narrative voice representing the family's fellow villagers. The author generally describes the settings with words that seem to come straight from the characters' mouths. This choral voice, Leo Spitzer states, constitutes "the originality of Verga's technique in *I Malavoglia* [...] in the *systematic* filtering of his narration of an entire novel, from the first to the last chapter, through a chorus of semi–realistic working–class speakers" (1956, p. 45).

The sea between metaphor and reality

As a result of this stylistic choice, the descriptions of the sea and of the events that unfold in this space are rarely made directly by the author, but are deduced from the words of the characters, both main and secondary.

The main events that constitute the defining moments of the story all unfold at sea: the sea represents the Malavoglias' main livelihood, with the identity of the main characters closely bound to the sea and to fishing. This is especially true for Padron 'Ntoni, who until his dying day will believe that the road to salvation for his family is through fishing because although it is a humble activity, it is dignified by the family's independence as owners of their fishing boat.

Thus, for the Malavoglia family, the sea assumes the value of an identity space. The author does not describe the other fishermen of Aci Trezza's relationship with the sea, only briefly mentioning their presence, for example telling how before dawn they could be heard going from house to house to rouse their companions on their way to the boats (p. 184) and how from the shore one could often see white sails in the distance "*che sembravano tanti gabbiani al sole*" (which looked like lots of seagulls in the sunshine) (Verga, 1881/1978, p. 166). It is thus a uniquely familiar identitary space, which in Verga's mind becomes the space that symbolises purity, closeness to a state of nature, while the village, and above all the vast world beyond, comes to symbolise a society threatened by utilitarian values and by interference from the modern state. Amid this changing world is an untouched island, a safe haven above all for the female members of the family, but where the menfolk also find respite from the toils of work: it is the Malavoglia family home, known as the house by the medlar tree, which represents a stronghold that defends their family values. The stubborn attempt to repurchase the house, a crucial element that is central to the novel's plot, represents the will to stand up to ever-advancing modernity, an attempt to protect the family as the natural social unit, and the chance to defend one's closeness to nature.

The events in which the Malavoglias are caught up spill over into the life of the village where they live, characterised by a well-developed network of economic and social relations that tend to involve entire families rather than mere individuals. Family interests, with on the one hand an almost morbid attachment to material possessions, which in some cases represent the only means of survival, and on the other hand a desire to maintain their social status, come before any individual interests. As a result, storylines that revolve around arranging marriages play a crucial role in the novel, as plans are hatched, shelved and re-hatched, with family interests trumping any consideration of individuals' feelings about such unions. This tapestry of relationships is complicated by the burgeoning intrusion of history, with allusions to a bigger world far beyond the lived experienced and collective imagination of the village, so that a national and European, if not global dimension encroaches on the local. This results in obligations imposed by the newly founded Kingdom of Italy, which introduce changes that have profound consequences on the lives of the protagonists[3] and which are the subject of heated debate by the inhabitants of Aci Trezza.

The vicissitudes in which the Malavoglias are embroiled are framed within this tapestry of social relations and are narrated chorally from the different points of view of those involved.

From a superficial reading, it might seem that the sea, despite being the space responsible for the Malavoglia family's tragedies and in which they seek salvation, is relegated to a marginal role in the story. This can be seen as in keeping with the Italian literary tradition, in which events at sea are rarely narrated, so much so that Raul Mordenti (2009) described it as a literature devoid of sea. Above all, this fact can be explained with the type of narrative used by Verga, entrusted to the words and points of view of the villagers of Aci Trezza, intent on the one hand at opposing and on the other at adopting the social models of the modern day. The sea, in contrast, is timeless, representing permanence, the opposite of modernity both because its elements are "the same as five hundred, a thousand, two thousand years ago – sun, sea, wind, sky, clouds, stars, rainbows, endless waves and still more waves; they will be the same in a thousand and ten thousand years' time" (Casati, 2022, p. 64), and because no precise boundaries can be traced at sea, meaning there can be no such thing as private property, which requires exact identification of the possession in question. One of the defining elements of modernity, private property is something to which a great many villagers aspire. Perhaps this is the key to interpreting the statement by one of the characters in the novel, according to whom "*chi ha roba in mare non ha nulla*" (he who owns stuff at sea owns nothing at all) (Verga, 1881/1978, p. 285), aside from indicating the precarious nature of items afloat on or in contact with the sea.

And yet the sea and seafaring activities provide a constant backdrop to the story, not only in descriptions, but also in the metaphors used by the characters in their language and the proverbs that pepper their everyday speech, used to reinforce their opinions, as much as they are seen to derive from incontrovertible wisdom. More than anyone else, it is Padron 'Ntoni who, Verga tells us "*sapeva certi motti e proverbi che aveva sentito dagli antichi 'perché il motto degli antichi mai menti*'" (knew certain sayings and proverbs of olden times "because the saying of the ancients never lie") (Verga, 1881/1978, p. 113) that is in the habit of using certain proverbs to back up his convictions and make sense of the world.

Looking closely at only those proverbs linked to the sea, the life of sea creatures and seafaring activities, we can see how they are used to interpret warning signs of meteorological conditions: "*scirocco chiaro e tramontana scura, mettiti in mare senza paura*" (light sirocco and dark tramontane, safely to sea once again) (Verga, 1881/1978, p. 124), "*quando il sole si corica insaccato si aspetta il vento di ponente*" (when the sun sets behind clouds, expect a west wind) (Verga, 1881/1978, p. 128), "*mare crespo, vento fresco*" (white horses at sea, a fresh wind there'll be) (Verga, 1881/1978, p. 128), "*mare bianco, scirocco in campo*" (white sea, sirocco there'll be) (Verga, 1881/1978, p. 276), "*quando la luna è rossa fa vento, quando è chiara vuol dir sereno, quando è pallida pioverà*" (a red moon signifies wind, a clear moon means fair weather, a pale moon warns of rain) (Verga, 1881/1978, p. 276). These proverbs are the product of careful and prolonged observation of the natural environment, and the desire to interpret its signs. Sayings linked to the sea and seafaring activities are also used to back up individuals' ideas as well as widespread beliefs on how society should be properly ordered and on the fate that awaits each member of society based on that order, which is nevertheless thrown into turmoil by modernity. So, for example, the need for a guide is summed up in the statement that "*senza pilota la barca non cammina*" (without a pilot a boat goes nowhere) (Verga, 1881/1978, p. 113), while the saying: "*chi è nato pesce il mare lo aspetta*" (he who is born a fish will find the sea awaits him) (Verga, 1881/1978, p. 184) alludes to the inescapability of fate, the same meaning conveyed by the proverb "*i pesci del mare son destinati a chi se l'ha da mangiare*" (the fish in the sea are destined for those who must eat them) (Verga, 1881/1978, p. 270), "*il buon pilota si vede nelle burrasche*" (storms reveal good pilots) (Verga, 1881/1978, p. 274) implies that the trials of life reveal people's abilities, while "*a nave rotta ogni vento è contrario*" (all winds blow against a broken ship) (Verga, 1881/1978, p. 259) is a reminder that problems seem to pile up and besiege those already in difficulty, "*per menare il remo bisogna che le cinque dita della mano si aiutino l'un l'altro*" (to pull the oar, all five fingers must help each other) (Verga, 1881/1978, p. 307) means that to succeed in an undertaking, group solidarity is essential, and "*per un pescatore si perde la barca*" (for one fisherman, a boat is lost) (Verga, 1881/1978, p. 353) means that the breakdown of group solidarity, even with the loss of a single member, can lead to the ruin of the whole group. One of these proverbs, which by no coincidence is repeated several times in the novel, is used to convey how harsh and dangerous working conditions are for sailors, and at the same time how their fate is linked to the sea "*il mare è amaro ed il marinaro muore in mare*" (the sea is bitter and sailors die at sea) (Verga, 1881/1978, p. 145, 175, and 187). In each case the proverbs are used to allude to some presumed wisdom legitimised by virtue of its antiquity, pointing to a desire to find unwavering guidelines on behaviour in order to interpret life, even though in that changing society their rules are increasingly inadequate.

Nautical metaphors pepper the whole novel and the author has them delivered not only by characters who work at sea, but also by others who carry out activities unrelated to this space, which suggests that the entire community of Aci Trezza has strong cultural ties to the sea, regardless of any economic ties. So, for example, to describe 'Ntoni's greed, we are told that "*è fatto come i merluzzi, che abboccherebbero un chiodo arrugginito*" (he's like cod, which would swallow a rusty nail used as bait) (Verga, 1881/1978, p. 119), while the Malavoglia family is described as "*quella barca della casa del nespolo*" (that boat of the house by the medlar tree) (Verga, 1881/1978, p. 121), the pain felt by Padron 'Ntoni over the loss of his son Bastianazzo is described as "*una grossa spina [...] che ci aveva in cuore, come se lo rosicasse un pescecane*" (a big thorn [...] sticking in his heart, as if a shark was gnawing at him) (Verga, 1881/1978, p. 162), to symbolise a person's generosity or courage, they are said to have "*il cuore più grande del mare*" (a heart bigger than the sea) (Verga, 1881/1978, p. 179 and 273), to describe the anger that spread through the village following news of the introduction of another tax, we are told that "*la bile andava gonfiandosi come l'onda del mare*" (bile swelled up like the waves of the sea) (Verga, 1881/1978, p. 212), the feeling of being lost is conveyed with the proverbial "*sentirsi un pesce fuori dell'acqua*" (to feel like a fish out of water) (Verga, 1881/1978, p. 330 and 332), the village inn, because of the comings and goings, is compared to "*un porto di mare*" (a sea port) (Verga, 1881/1978, p. 385) and so on.

Considering that the main characters work at sea, that the entire village is steeped in nautical culture, and that the event that triggers the Malavoglias' tribulations occurs at sea, the author provides relatively few direct descriptions of this space, in deference to Verga's desire to conceal the role of the narrator as much as possible.

Thus, despite Monica Farnetti's observation that "tempests or squalls, with rough seas accompanied by bad weather, storms and downpours, are one of the main topical features in maritime literature" (1996, p. 175), for the sake of a choral narrative voice, Verga chooses not to describe the storm in which the Malavoglias' boat sinks, taking with it not only its valuable cargo but also Bastianazzo and his fishing companion, dragging the family into a downward spiral of negative events that culminate in terrible misfortune. The reader learns about the disaster through the description of the effects of the wind in the village which "*s'era messo a fare il diavolo, come se sul tetto ci fossero tutti i gatti del paese*" (began howling as if every cat in the village were on the roof) (Verga, 1881/1978, p. 146), through the worried thoughts of Padron 'Ntoni and, following the disaster, through the voices and considerations of the villagers of Aci Trezza.

The descriptions of the sea and seafaring activities seem to be used to pit the feelings of human beings who are focused on the difficult task of improving their social condition or adapting to the demands of modernity, against the eternity and cold indifference of the natural elements. The sea is often anthropomorphised, but the words which the author has the characters speak, or which could have been pronounced by the characters, do not hide the sea's essential indifference to the fate of humans. For example, on the day of

Bastianazzo's funeral, Verga describes the sea in the following terms: "*I vetri della chiesetta scintillavano, e il mare era liscio e lucente, talché non pareva più quello che gli aveva rubato il marito alla Longa*" (The windows of the little church glinted in the sun, and the sea was mirror-calm, no longer bearing any resemblance to that which had snatched la Longa's husband) (Verga, 1881/1978, p. 157). Elsewhere the sea is described as someone "*che russava lì vicino, in fondo alla straduccia, e ogni tanto sbuffava, come uno che si volti e rivolti pel letto*" (snoring nearby, at the end of the little road, who puffed every now and again, like someone tossing and turning in bed) (Verga, 1881/1978, p. 143) and the sea is also described as "*il vecchio brontolone*" (an old grump) (Verga, 1881/1978, p. 153). At points, descriptions of the sea appear to paint idyllic scenes, but even though the midnight-black sea "*si specchiavano le stelle e [...] russava lento sul greto*" (reflected the stars and [...] murmured softly over the shingle) (Verga, 1881/1978, p. 187), and it is described as "*liscio e turchino*" (smooth and turquoise) (Verga, 1881/1978, p. 166), "*verde come l'erba*" (as green as grass) (Verga, 1881/1978, p. 273) and "*tutto d'oro*" (pure gold) (Verga, 1881/1978, p. 187), it remains indifferent to human troubles.

It is yet another storm that seals the fate of the Malavoglias. Indeed, once the wreck of the family fishing boat is recovered and clumsily repaired, it is then given the chance to put to sea again to fish and thus hopefully make the money needed to settle the family's debt and buy back the house by the medlar tree. In this case the sea is described in an entirely different tone: Verga writes that "*il mare era del color della Sciara*[4]*, sebbene il sole non fosse ancora tramontato, e di tratto in tratto bolliva tutto intorno come una pentola*" (the sea was the colour of black lava flows, even though the sun hadn't set yet, and every now and again it hissed and bubbled like a pot) (Verga, 1881/1978, p. 277), "*le onde, quando passavano vicino alla Provvidenza*[5]*, luccicavano come avessero gli occhi e volessero mangiarsela; e nessuno osava dire più una parola, in mezzo al mare che muggiva fin dove c'era acqua*" (when the waves rolled past the Provvidenza, they shone as if they had eyes and wanted to devour it; and nobody dared utter a word in the presence of a sea that moaned as far as the water stretched) (Verga, 1881/1978, p. 278), "*la Provvidenza cominciò a balzare sulla cima delle onde, piegata da un lato come un uccello ferito*" (the Provvidenza began to dance on the crests of the waves, leaning to one side like a wounded bird) (Verga, 1881/1978, p. 281) and "*la notte era così nera che non si vedeva da un capo all'altro della Provvidenza*" (the night was so dark, it was impossible to see from one end of the Provvidenza to the other) (Verga, 1881/1978, p. 282). Although Monica Farnetti states that a raging sea "is the one of the most recurrent 'enemies' in the seafaring adventure model, suited to triggering religious sentiment in a character or, in contrast, the rebellious energy of pride, courage and defiance" (1996, p. 175), Padron 'Ntoni does not consider the sea to be his enemy; he believes that life at sea, for all its dangers and misfortunes, is a destiny that cannot be escaped. While some of his fellow villagers think it wise to cut ties with the sea and abandon fishing as a livelihood, a consideration voiced by of one of the neighbours after the *Provvidenza's* second shipwreck:

> *quel mestiere lì è fatto in tal modo, e si finisce col lasciarci la pelle. E una che mariti la figlia con gente di mare [...] un giorno o l'altro se la vede tornare a casa vedova, e con gli orfani per giunta* (that's the way with this job, and it ends up taking your life. And anyone who marries their daughter to a seafaring man [...] will sooner or later see her return home a widow, and with children, to boot).
>
> (Verga, 1881/1978, p. 284)

He knows that his misfortunes stem from the ill-fated idea to try to climb the social ladder, as his business venture selling lupins was a betrayal of the sea.

Padron 'Ntoni's relationship with the sea is not only linked to his acceptance of a tragic fate, but is interwoven with his love for that space, which for him can be considered a locus. Verga describes his return to fishing, following the first tragic shipwreck, in that instance as a hired worker since the "Provvidenza" was under repair, with these words: "*Ah! Esclamò stirandosi le braccia. È una bella cosa tornare a casa sua. Questa marina qui mi conosce – Già Padron 'Ntoni diceva sempre che un pesce fuori dell'acqua non sa starci, e chi è nato pesce il mare l'aspetta*" ("Ah," he exclaimed, stretching his arms. "It's a fine thing to come back home. This sea knows me. As Padron 'Ntoni always said, no fish can live out of water, and he who is born a fish will find the sea awaits him") (Verga, 1881/1978, p. 184).

This attachment to the sea is also revealed in the following passage which describes his behaviour when, recovering from the second shipwreck which has aged him physically and emotionally:

> *cominciò a stare sulle gambe, e lo portarono sulla riva, reggendolo sotto le ascelle, perché gli piaceva dormicchiare accoccolato sui sassi, in faccia alle barche, e diceva che l'odore dell'acqua salata gli faceva bene allo stomaco; e si divertiva a vedere le barche, e sentire com'era andata la giornata per questo e per quello* (once he was able to stand again they took him down to the shore, supporting him under the armpits, because he liked to doze on the pebble shore, facing the boats, and he said the smell of the salt water was good for his stomach; and he enjoyed seeing the boats, and hearing how the day had gone for this one and that one).
>
> (Verga, 1881/1978, p. 301)

For Padron 'Ntoni, the sea is therefore a space to which his fate and that of his family is tightly bound, and despite its tragic nature it is a fate that is not only accepted but considered the only one possible; rejecting that fate entails a concatenation of misfortunes, the loss of his identity, his independence, and his dignity. It is also a beloved space, a locus.

Verga juxtaposes the figure of Padron 'Ntoni with that of his homonymous grandson, corrupted by his experience of the big wide world encountered in Naples during his military service, which plants in his mind the idea that it might be possible to escape the miserable life of a fisherman. His aversion to that life lived in symbiosis with nature, following its ever-repeating rhythms and never contemplating any possible economic and social betterment, leads him to a doomed attempt to leave his family and the village and, when he returns, the feeling of failure coupled with a refusal to accept his fate leads him to alcoholism, trafficking in contraband goods, criminal activity and, ultimately, imprisonment. For Verga the consequences of refusing to accept one's fate are irreversible. On his release from prison, 'Ntoni returns to his village, to the house by the medlar tree, repurchased by his younger brother Alessi who, unlike him, has carried on the family tradition, and he finally realises what he has lost, that the death of Padron 'Ntoni, of his mother la Longa, and the loss of his younger sister Lia, are irremediable and that there is no longer a place for him in that world. Before leaving for good, he wanders around the village as if trying to absorb the images of that world now lost to him, which is symbolised by the sea:

> *Soltanto il mare gli brontolava la solita storia lì sotto, in mezzo ai farighoni, perché il mare non ha paese nemmeno lui, ed è di tutti quelli che lo stanno ad ascoltare, di qua e di là dove nasce e muore il sole, anzi ad Aci Trezza ha un modo tutto suo di brontolare, e si riconosce subito al gorgogliare che fa tra quegli scogli nei quali si rompe, par la voce di un amico* (Only the sea murmured its usual story, down there between the sea stacks, for the sea too has no country and belongs to all those who stop to listen to it, here or there, wherever the sun rises and sets, and at Aci Trezza it has its own unique way of grumbling, which one can recognise instantly from its gurgling among the rocks where it breaks, and it sounds like the familiar voice of a friend).
>
> (Verga, 1881/1978, p. 431)

Conclusions

Verga therefore tasks his novel with expressing his political and social beliefs, his idea that the irreversible advance of the modern world would alienate mankind from the state of nature, and that although it might improve the existential condition of the poorest members of society, it would entail the loss not only of their identity, but of that acceptance of their unhappy fate which alone could afford some kind of serenity.

The sea, therefore, is not the backdrop to the narrated events, it is a metaphor for purity now lost. Descriptions of the sea are not just narrative elements employed to frame the story, but tesserae that form the mosaic of Verghian ideology which, given the power of his storytelling, have influenced and continue to influence on the one hand values attributed to the sea, and on the other the way of viewing society.

Today in Italy modernity has imposed itself and with it consumerist and urban life models, and the sea and its landscapes have been subjected to profound commodification through exploitation for seaside and more generally tourist purposes. Thus, the stacks that in the novel's final pages represent the symbol of the stability of human destiny over time have become mere naturalistic attractions and the author's choice to set his work in Aci Trezza a resource for activating processes of tourist–cultural development (Marengo, 2016, p. 20). However, the sea remains a powerful metaphor for the impossibility of completely bending nature to human will, and for this reason, Verga's work can be read as a warning not to forget the role of nature in human affairs, which would have consequences not only on an environmental level but also on a social and ethical one.

Notes

1 Ecocritical analysis, which first emerged in literature, was later applied to other narrative genres including cinema, theatre, the visual arts, comic strips and so on.
2 To better understand this statement, let us briefly remind ourselves of the novel's plot. The business venture that sets the whole chain of events in motion is the purchase on credit of a batch of lupins, to be taken using the family fishing boat, named *Provvidenza* (Providence), to Riposto, around 20 kilometres from Aci Trezza, and sold for a profit. Tragedy strikes and the ship is lost at sea along with its cargo of lupins. The family not only loses Bastianazzo, who would one day have taken over from his father as head of the family, but is also forced to cede the family home to the creditor in order extinguish the debt racked up for the lupins. The sale of the property could have been avoided given that it was part of the dowry of Maruzza, Padron 'Ntoni's daughter–in–law, but as he was determined to protect the Malavoglia family's reputation for honesty, he sold it anyway, irreversibly worsening the family's situation.
3 For example, the introduction of compulsory military service is what first forces young 'Ntoni to leave his family, and then obliges Luca to fight in the battle of Lissa where he is killed; the levying of new taxes fuels trade in contraband goods, again involving 'Ntoni and leading to his imprisonment.
4 In Sicilian dialect, this term indicates areas where lava flows have solidified, leaving dark coloured rocks or scoria.
5 It is the name of the Malavoglia's boat.

References

Brosseau, M. (2022). *Tableau de la géographie littéraire*. Presses Universitaires de Pau et des Pays de l'Adour.
Carnazzi, G. (1978). Introduzione. In G. Verga (Ed.), *I Malavoglia*. Rizzoli (original work published 1881).
Casati, R. (2022). *Oceano. Una navigazione filosofica*. Einaudi.
dell'Agnese, E. (2021). *Ecocritical geopolitics. Popular culture and environmental discourse*. Routledge.
Dematteis, G. (1985). *Le metafore della terra. La geografia umana tra mito e scienza*. Feltrinelli.
Devoto, G. (1954). I "piani del discorso" in due capitoli dei "Malavoglia". *Bollettino del Centro di Studi filologici e linguistici siciliani, II*, 5–13.
Farnetti, M. (1996). *Il romanzo del mare. Morfologia e storia della narrativa marinara*. Le Lettere.
Marengo, M. (2016). *Geografia e letteratura. Piccolo manuale d'uso*. Pàtron.
Matvejević, P. (1993). *Mediterraneo. Un nuovo breviario*. Garzanti.

Mordenti, R. (2009). La letteratura senza mare. In C. Cazalé Bérard, S. Gambino-Longo, & P. Girard (dir.), *La mer dans la littérature italienne* (pp. 25–46). Presses Universitaires de Paris Ouest.
Spinazzola, V. (1977). Legge del lavoro e legge dell'onore nei 'Malavoglia'. In V. Spinazzola (Eds), *Verismo e positivismo*. Garzanti.
Spitzer, L. (1956). L'originalità della narrazione nei "Malavoglia". *Belfagor*, *11*(1), 37–53.
Turco, A. (2010). *Configurazioni della territorialità*. Franco Angeli.
Verga, G. (1978). *I Malavoglia*, Milano, Rizzoli (Original work published 1881).
Wright, J. K. (1947). Terrae incognitae: The place of imagination in geography. *Annals of the Association of American Geographers*, *37*(1), 1–15. https://doi.org/10.2307/2561211

17 The Italian countryside between rurality and nature

Wandering around Cesare Pavese's hills and novels

Dino Gavinelli and Giacomo Zanolin

Introduction

Literary representations do not necessarily imitate reality, rather they can be understood as practices through which socially shared meanings are constructed to give significance to what is concrete and experiential. The idea of nature is no exception in this respect and can be considered as a representation, co-constructed and re-constructed through social and cultural practices (Berque, 1996; Castree & Braun, 2001). For this reason, it may be useful to study the meaning of the idea of nature from a geo-literary approach when investigating ecological dynamics from a cultural perspective.

In line with these theoretical and methodological approaches, this chapter presents a reflection on the ambiguous relationship between rurality and nature in 20th literary representations of the Italian countryside. The rural world is the protagonist of many novels from the post-war period onwards and has contributed to the construction of an idyllic collective imaginary in which rurality and nature tend to merge.

Starting from a critical reading of some of Cesare Pavese's novels, we propose to reflect on the paradox of an Italian rurality represented both as the legacy of a past that was founded on human labour and as the expression of an idea of nature unaffected by human action. As we will try to demonstrate, these texts bring to the fore the constructive role played by human beings in the long course of history in designing rural landscapes, cooperating with non-human living beings in generating ecosystem dynamics.

The critical geographical reading of Pavese's rural idyll allows us to call attention the intrinsic rurality of all that is usually considered natural in Italy. The aim is therefore to understand this process of social construction of an idea of nature that is related to rural Italian landscapes, strictly connected to nation-building and nation-branding processes.

Theoretical insights

This paper is based on a geo-literary approach, that reflects on the symbolic meanings attributed to geographical space that cannot be considered as an objective entity, but as a subject that is socially constructed and interacts with

DOI: 10.4324/9781003465164-22

all living beings (obviously including humans). The potential for fruitful dialogue between geography and literature has been extensively studied internationally for several decades (Brosseau, 1994, 2022; Gavinelli and Marengo, 2021; Hones, 2022; Marengo, 2022). In such an intense debate, we can trace the distinction between a literary geography, which favours analyses that are internal to the text, aimed at analyzing the representations of spatial dimensions, and a geography of literature aimed at understanding the relationships between literary works and the territorial contexts that produced them (Brosseau & Cambron, 2003).

The literary text cannot be considered solely as a source for geographical studies, but also as a subject through which a process of signification is initiated or reiterated, aimed at constructing a socially shared perspective on space, which takes shape in the dimension of landscape (Castiglioni, 2022). Landscape is a central concept in contemporary society, as highlighted by several scholars who have used expressions such as "omnilandscape" (Jakob, 2009) or "landscape thinking" (Berque, 2016) precisely to express the pervasiveness of this concept as a filter through which we experience the world and as an elucidatory tool through which we interpret its dynamics. It is evident in this context that the literary text emerges as one of the subjects that, by constructing representations of the landscape, activate a process of social construction of reality that can generate a shared imaginary about specific spatial contexts (Berger & Luckmann, 1969; Bourdieu, 1979).

The meaning of landscape in contemporary society is complex and stems from its intrinsic polysemic nature. Studies on landscape cannot therefore be limited to its material dimension, rather they must be carried out seeking a balance between objectivity and subjectivity, rationality and emotionality. In this regard, Alain Roger (1997) spoke of "artialisation", to emphasize how landscape is generated from the construction of a discourse mediated by art. One of the privileged subjects within these discourses has long been the rural landscape, and it is precisely this concept that represents the most important theoretical polarity around which this chapter develops. To give a succinct definition of rural landscape, we must first clarify how we understand rurality, as its meaning is much less intuitive than we might think at first glance (Woods, 2005). The idea that rural should be understood as everything antithetical to the urban is now outdated, in light of the awareness that urbanization and ruralization are two complementary outcomes of the process of territorialization. Similarly, rural should not be understood as synonymous with agrarian, as in contemporary times rural landscapes undeniably characterize regions in which primary activities do not necessarily play a prominent economic role. Today, therefore, the term rural expresses the hybridization of territorial processes rather than the distinction between rigid categories (Cloke, 2006). Since the material forms of the rural landscape express this complexity, it is evident that its discursive form contains a multiplicity of meanings that are difficult to synthesize. Among these, regarding the reflections addressed in this chapter, it is worth noting the ability of the rural landscape to evoke the idea of rural idyll, "one of the most powerful and enduring ideas about the rural" (Woods, 2005, p. 21). This idealistic vision of the rural landscape is useful to affirm that

> the countryside is [...] a peaceful and virtuous landscape, opposed to the busy and problematic city, which has led to the idealization of rural areas and a nostalgic vision [...] Ironically [...] in contemporary times, when cities have become dominant and rural areas have suffered intensive processes of depopulation as a result of the persistent rural flight, the rural idyll has (re–) emerged strongly and become widespread.
>
> (Paül et al., 2020, p. 248)

Therefore, the rural landscape configures itself as an elsewhere, an almost mythological expression of a space antithetical to that where everyday life unfolds, that readers look upon with nostalgia, as if it were the site of a lost paradise. Tourism obviously strongly exploits this evocative capacity of the rural landscape, in order to guide tourists from cities to choose rural destinations for their vacations. Emblematic of this process is the symbolization of the Tuscan hill landscape as a synthesis of the Italian rural landscape from some cinematographic works, such as *The English Patient* or *Under the Tuscan sun* but above all through the British TV series *Summer's Lease*, based on the eponymous novel by John Mortimer, in which the tourist myth of *Chiantishire* is founded (Meloni, 2023). The elsewhere constructed through discourses on the rural landscape is not simply a spatial elsewhere, but also a temporal elsewhere, which projects the tourist into "an ancestral, almost magical temporal dimension, whose charm lies in its intrinsic diversity compared to the time dictated by Taylorism and Fordist work organization that characterize modernity"[1] (Aime & Papotti, 2012, p. 39).

To conclude, it is important to emphasize that these idyllic and stereotyped visions of the rural landscape assume concrete relevance not only as potential economic resources available to the tourism industry, but also, in many cases, acting as political tools for nation-building (Woods, 2005). The strong relationship between nation-building and the rural landscape has been well described by Depraz (2008). The French geographer highlighted not only the strong connection between rurality and nature, which often tends to merge when described as stereotyped antitheses of urbanity, but also the ability of many rural landscapes (erroneously described as natural) to configure themselves as territorial identities of some sort of haven. These refuges are described as spaces preserved from the process of civilization and time, where the archetypes upon which national principles are based arise. In the case of Italy, as already mentioned with reference to Tuscany, hilly areas take on the dimension of "iconemes" (Turri et al., 2001) that rise to the role of identity representations. Some hilly regions are at the center of this process, as the already mentioned Chiantishire or the recently included Umbria and Marche's inner areas, composing an idealized vision of the hills of central Italy in the collective imagination. In Northern Italy, this role is played, with increasing importance since the 1980s, by the region known as Langhe–Roero–Monferrato, a vast system of morainic hills in southern Piedmont recognized in 2014 as a cultural landscape in the UNESCO World Heritage List (Pettenati, 2019). This peculiar landscape has been materially shaped over history by various agricultural activities;

towards the second half of the 20th century, it has found an increasingly evident role in viticulture and therefore in the production of globally renowned high-quality wines. Since the first years of the 21st century, a process of heritagization has started to increase the attractivity of the region, focusing on viticulture landscapes. This paper is based on the belief that the literary work of Cesare Pavese (1908–1950) contributed[2] to build a specific idea about those landscapes, based on the nostalgia for the rural age that predates contemporaneity and so to the construction of an imaginary related to these areas linked to rurality, on which this chapter focuses.

Cesare Pavese between realism and introspection

Cesare Pavese, born in Santo Stefano Belbo (Cuneo, Italy) in 1908, was one of the most important Italian writers, poets, translators, and literary critics of the first half of the 20th century. Having graduated with a thesis on Walt Whitman and translated numerous British and North American authors including Daniel Defoe, Charles Dickens, John Steinbeck, and Herman Melville, he adopted a Conradian narrative technique, but was also influenced by the works of Ernest Hemingway, which contributed to shaping his early writings from the 1930s (Gigliucci, 2001). The first novel through which he attempted to transplant the modes of North American narrative into Italy is entitled *Paesi tuoi* [*The harvesters*] (1941), where the Langhe region is somewhat transformed into a provincial Far West; the same vision is found in another work entitled *La casa in collina* [*The house on the hill*] (1948):

> Behind Le Fontane there were vineyards and fields with crops, and I went there often, to gather herbs and mosses in sheltered little glades, an ol hobby from when I'd studied natural science. I always preferred ploughed fields to houses and gardens, and the edges of fields where the wild takes over.
>
> (Pavese, 1948/2021a, p. 28)

Despite its realistic approach, that literary critics associated to *Verismo*, some key themes of Pavese's poetics emerge in this novel: characterized by a profound introspective charge, but constructed from a narrative that was strongly anchored to the context in which the story unfolds.

Like many other Italian authors of the first half of the 20th century,[3] Pavese was influenced by reading Sigmund Freud's texts on psychoanalysis (Marchese, 2014). Nevertheless, his writings were not limited to representing his inner space, rather they aimed to express the author's particular point of view on the transformations that were characterizing the hilly regions of southern Piedmont during his lifetime, when the process of abandonment and depopulation in favour of urbanization was beginning in the polarizing nearby centre of Turin. In Pavese's literature, therefore, landscape takes shape as a sort of mirror on which he builds a strong critique of the socio-territorial processes that were underway, projecting in it the personal difficulties that accompanied his life, leading to his suicide in 1950, at the age of 42 (Vaccaneo, 2020).

Life, understood as a "tiring profession", is one of the two main protagonists in Cesare Pavese's work, which, however, is not directly represented through his characters, but is indirectly portrayed through the description of a decaying landscape in the Langhe:

> The hill is no longer cultivated up here. There are the ferns
> and the bare rock and the infertility.
> Work no longer serves any purpose here. The summit is burnt
> and the only freshness is the breath. The great effort
> is to climb up here: the hermit came here once
> and has stayed ever since to restore his strength.[4]
>
> (Pavese, 1936/2001a)

> I didn't think that even in the country on Sunday mornings there would be a difference, but you can really see that working offends the air, because that morning everything was cooler and more peaceful.[5]
>
> (Pavese, 1941/2000, p. 61)

Pavese's narrative work is characterized by the use of a dynamic language, with many dialogues, and the adoption of cinematic narrative techniques (including numerous flashbacks). The use of adventurous or mythical characters sets him apart from *Verismo*,[6] while still proposing an epic representation of everyday life and regression to the origins of the hilly regions according to the *veristic* taste:

> I told him that in my day the valley here was grader, there were people travelling around in carriage and men with waist-coats and gold chains, and the village women and Station women carried sunshades. I told him about the parties they threw and the festivals – wedding, baptisms, Madonna processions – and people came from all over, way up on the hills, musicians, hunters, village mayors. There were houses – villas, like Villa Nido up on the hill above Canelli – with rooms big enough to sleep fifteen or twenty, like at the Hotel Angelo, where people would eat and play music all day. And we'd have parties in the farms too, us kids, playing *settimana* in summertime, or spinning on the ice in winter. In *settimana* you ha to stand on one leg, the way Cinto was standing, and hop across rows of pebbles without eve touching the pebbles. After the grape harvest the hunters would set off into the hills from Gaminella, San Grato or Camo, coming back covered in mud, dead tired, but loaded with partridges, hares and game. We'd see them passing by from the house and later you'd hear people partying in the village into the night, and in Villa Nido over there – you used to be able to see, because those trees weren't there then – all the windows would be lit up, like there was a fire, and you'd see shadows of people flitting across them till morning.
>
> (Pavese, 1950/2021b, pp. 28–29)

Pavese thus contributed in bringing to Italy a vitalistic and activistic alternative to the narrative based only on introspection. He can be described, at the same time, as the architect of the realistic renewal of Italian narrative along with Elio Vittorini (1908–1966), and as an author that embodied decadent values that strongly set him apart from true realism. It is this peculiar complexity that enhances the symbolic charge of Pavese's work. His entire opus as a poet and narrator is thus an interweaving of realism and decadence. Pavese's existential effort lies precisely in the desire to exhaust decadence within itself and to give his sufferings a realistic voice through the representations of the landscape.

Effective in this regard is his poetics of myth or of always seeing for a second time, which refers to the representation of rural and urban worlds as temporal and spatial experiences: projecting the transition from childhood and adolescence to adulthood, experienced as a process leading to the loss of existential values of fundamental importance but which are re-elaborated and presented in an almost archetypal form (Pavese, 2021c). Always seeing for a second time consists precisely in rediscovering and giving rational clarity or meaning to childhood myths in light of the exhausting experiences of adulthood. The possibility of freely shaping one's existence is entirely illusory, and the individual is conditioned by its childhood vision. The contrast between childhood and maturity is therefore reflected on ~~a~~ historical, geographical, and social levels (childhood–primitive age–countryside versus maturity–civilization–city). Pavese is rationally aligned with the latter terms (maturity–civilization–city); he knows that individuals must become adults, that history is progress and civilization, that the city awakens individuals and makes them more open to life, human problems, social and territorial duties, but emotionally and ideologically, he remains attached to the former terms of the contrast (childhood–primitive age–countryside) (Gigliucci, 2001).

Explanatory of this approach is certainly the novel *La luna e i falò* [*The Moon and the Bonfires*] (1950), right from the first pages:

> Who knows what flesh I'm from? I've travelled the world enough to know that one is as good as the next, which is why you weary of travelling and want to put down roots, make a piece of land your own and be part of it, so your flesh can count and last something more than a turn of the seasons.
> (Pavese, 1950/2021b, p. 5)

In practical life, he suffers this awareness: he lives in the city, is involved in politics, is a successful man, but he perpetually feels uprooted and never entirely at ease: disorder, alcohol, sexuality, the return to his childhood hills, and, above all, writing, are the ways in which he attempts to react to his uprooting (Mondo, 2006).

Pavese feels that it is his duty to be mature as a man and as a citizen; he wants to participate in history and in the world, so he commits himself, but he remains inexorably nostalgic for his childhood. At the origin of Pavese's symbolic characters, there is this lack of sentimental education, there is the unyielding contrast between reason and mitigating childish fantasy. And, it is precisely such a dualism that prevents him from creating true human characters and rather leads him to attribute a protagonist role to elements of the landscape: a tree, the sound of a cicada, a cornfield, a vineyard, a river, a bonfire, and so on.

The landscape and Cesare Pavese

As mentioned, the second protagonist in Cesare Pavese's work is undoubtedly the landscape, which is both a reflection of the author's existential crisis and a representation of the transformation of territorial processes taking place in Italy in the early 20th century. The result is a plausible (and therefore believable) representation of space that is not realistic but constructed with the purpose of expressing his aforementioned subjective discomfort. The outcome is, in a way, an objectification of the author's subjectivity through the figure of the landscape, leading to the construction of an idealized vision of the rural landscape of the hills of southern Piedmont. In this idealized vision, the hilly landscape becomes an expression of a rurality that configures itself as both a spatial and temporal elsewhere, projecting onto an idealized past the possibility of a better life that seems impossible in the daily life experienced by Pavese in the city.

This particular perspective on the rural landscape can be found both in poetry and prose:

> It had been the same in the past, only dotted with lights, people living peacefully in either houses, rest and jollity. Even now sometimes a voice would burst out, you'd hear laughing in the distance, but the thick darkness weighed heavily, covered everything up, and the earth had turned wild again, out there alone, the way I knew it as a boy.
>
> (Pavese, 1948/2021a, pp. 5–6)

The peasant world and the Langhe's landscape are a sort of microcosm through which Pavese constructs a mythology of the hills as a place of lost time. Pavese portrays essential aspects of the reality of his time in his stories and novels with a passionate and bitter tone, that sometimes includes an implicit moral judgement. Pavese oscillates between chronicle, description, and fantastic–memory reconstruction linked to his myths, especially those of time, seasons (among which summer emerges as the favourite, the era of amorous discoveries and revelations), hills, and nature. These are the themes that characterize *La spiaggia* [*The Beach*] (1942), *La bella estate* [*The beautiful Summer*] (1949/2018), and *Il diavolo sulle colline* [*The Devil in the Hills*] (1949).[7] The latter, in particular, is a grotesque dance of "possessed citizens" in the primordial, wholesome, and harsh world of the hills. The protagonists are the corrupted children of millionaires who pose as extravagant existentialists. Here, Pavese highlights truly nauseating and unhealthy aspects of parts of urban society, presenting citizens "possessed" by a materialistic conception of life typical of an era, condemning the generation of capitalists who have placed wealth at the centre of life's meaning. The devils on the hill are therefore the furthest thing from nature, from the world of the hills, and from the peasants.

Throughout this, it is evident that the temporal elsewhere represented by the peasant world also reflects a natural elsewhere, in which echoes of Whitman's poetry resonate and indirectly lead, in the Italian context, to a mystification of fundamental importance from the standpoint of a social construction of the concept of nature. Rurality becomes a place of primordial idyll in which paradoxically humans are perceived as intruders, becoming the cause of the degradation of what their ancestors had wisely built. These ancestors are considered as distinct entities from contemporaries, much like Native Americans who, in constructing the rhetoric of wilderness, have been excluded (symbolically and materially) from the historical processes of co-construction of the North American landscape, that they inhabited as protagonists for centuries. The existential need to immerse in primordial nature is particularly evident in *La casa in collina* [*The House on the Hill*], as can be inferred from the following exemplary quotes:

> The truth is only I went to enjoy the feeling of being near the woods, about to head off up there. In the savage stillness of the July sun, the familiar faces round the familiar table and the drawn-out goodbyes filled my heart with pleasure.
>
> (Pavese, 1948/2021a, p. 30)

> Climbing back up to the hill, I sensed how precarious this refuge had become. There was a feeling of expectation in the silent woods. The sky was empty too. I wished I was a root, or a worm, and could sink down into the ground.
>
> (Pavese, 1948/2021a, pp. 60–61)

In *La luna e i falò* [*The Moon and the Bonfires*], instead, the call to the ancient theme of the return journey, the Greek "nostos", emerges (Mencarini, 2013). The main character is a migrant farmer who speaks in the first person; this allows Pavese (drawing on Giovanni Battista Vico) to attribute to the farmer, the primitive, the savage, and more generally to the natural, not only a nostalgia for the ancestral rural world and for the countryside of his childhood but also a broader aspiration for a primitive and simple life. A life, thus, that possesses characteristics that are found in Latin and bucolic literature in contrast to the Etruscan element, an expression of a mentality and a form of urban, refined, and cultured life. A similar opposition is recovered by Pavese from the antithetical pair *rus–ars* present in Horace's first epistle to Augustus.

> It was odd how everything had changed but was still the same. Not a single one of the old vines remained, not a single animal; what had been meadow was stubble, and what had been stubble was vineyard, people

> were gone, grown up, died; roots torn up in ladslips, carried off by the Belbo – yet looking around, at the hillside that was Gaminella, the tracks on the Salto hill in the distance, the farms, the wells, the voices, the hoes, everything was as it always had been, everything had the same smell, taste, colour of the past.
>
> (Pavese, 1950/2021b, p. 28)

The hilly landscape that is the protagonist of Pavese's work is therefore not merely *veristic* or simply realistic but lyrical. The representation of the landscape cannot be solely descriptive but is functional to the construction of a metaphorical framework that stimulates participation, involves characters, witnesses, shared memories, thus prompting feelings derived from obscure semblances and memory. It is a geographical landscape but also a presence untouched by time and towering over its relentless flow. The landscape becomes an ancestral presence, a sort of fixed and eternal figure, as it were, for the animistic vision and mythical sensitivity of the ancients, the sun, the moon, the stars, and all those forces that make life possible but at the same time can threaten the lives of human beings on Earth.

Conclusions

As a result of this paper, Pavese's texts reveal an emotional narrative deeply immersed in a geographical landscape. The described landscape elements are realistic and seem to form part of an objective picture. Readers trust the author because of this, and the lyrical and symbolic meanings he seeks to convey are not perceived as a purely subjective point of view. Consequently, the effect on the readers' imagination is powerful: they feel as though they are receiving information about both the material and immaterial landscape. In reality, however, Pavese's emotional and epic narrative is rooted in the spatial dichotomy between the urban and rural, as well as in memory and nostalgia for the past.

The rural and hilly landscapes of the Langhe region he describes are therefore explored in their hidden depths and become idyllic spaces. These spaces are depicted as a "natural" antithesis to the urban world, with its competitive, chaotic, and alienating rhythms. Nonetheless, their nature is inherently "rural", shaped by human labour in ecosystems over centuries of history. The rural landscape thus becomes a kind of metonym for the suffering of those who struggle to live in a contemporary world defined by an ever-widening gap between human life and ecological dynamics. It is configured as a space–time elsewhere, a place where the possibility of living in peace with oneself and with society is envisioned.

In this rural landscape, an idea of nature emerges – one in which human beings can rediscover their lost integrity by distancing themselves from the contemporary world, which generates suffering and a pervasive sense of inadequacy. This is a condition the author shares with his readers to create his distinctive idea of nature. This idea becomes a socially shared construct, thanks to his literary work.

In conclusion, we may emphasize that Cesare Pavese's poetics effectively stimulate reflections on the interaction between the self and the world (Messina, 2022), proposing a melancholic attitude towards a rural idyll set in an imagined past, where the relationship between human beings and nature was more authentic. This relationship, Pavese suggests, allowed for the inner peace he sought but never fully managed to attain.

Notes

1 Our translation. Italian version: "una dimensione temporale ancestrale, quasi magica, il cui fascino affonda nella sua intrinseca diversità rispetto al tempo scandito dal taylorismo e dall'organizzazione del lavoro fordista che caratterizzano la modernità".
2 Other authors such as Beppe Fenoglio have been similarly relevant, although this paper is based only on the works by Cesare Pavese because of his fundamental cultural role in Italian literature on hilly landscapes.
3 Such as Luigi Pirandello (1867–1936) or Italo Svevo (1861–1928).
4 Our translation. Italian version: "Non è più coltivata quassù la collina. Ci sono le felci/e la roccia scoperta e la sterilità./Qui il lavoro non serve più a niente. La vetta è bruciata/e la sola freschezza è il respiro. La grande fatica/è salire quassù: l'eremita ci venne una volta/e da allora è restato a rifarsi le forze".
5 Our translation. Italian version: "Non credevo che anche in campagna il mattino della domenica ci fosse una differenza, ma si vede proprio che lavorare offende l'aria, perché quella mattina era tutto più fresco e più tranquillo".
6 Italian literary movement of the second half of the 19th century, based on the desire to represent the harshness of the social context of the time.
7 For the English version of *The Beach* and *The Devil in the Hills*, see Pavese, 2001b.

References

Aime, M., & Papotti, D. (2012). *L'altro e l'altrove. Antropologia, geografia e turismo*. Einaudi.

Berger, P. L., & Luckmann, T. (1969). *La realtà come costruzione sociale*. Il Mulino.

Berque, A. (1996). *Etre humains sur la Terre. Principes d'éthiques de l'écoumène*. Gallimard.

Berque, A. (2016). *La Pensée paysagère*. Aux éditions éoliennes.

Bourdieu, P. (1979). *La Distinction. Critique sociale du jugement*. Les Éditions de Minuit.

Brosseau, M. (1994). Geography's literature. *Progress in Human Geography*, *18*(30), 333–353.

Brosseau, M. (2022). *Tableau de géographie littéraire*. Presses universitaires de Pau et des Pays de l'Adour.

Brosseau, M., & Cambron, M. (2003). Entre géographie et littérature: frontières et perspectives dialogiques. *Recherches sociographiques*, *44*(3), 525–547.

Castiglioni, B. (2022). *Paesaggio e società. Una prospettiva geografica*. Carocci.

Castree, N., & Braun, B. (2001). *Social nature: Theory, practices, and politics*. Wiley-Blackwell.

Cloke, P. (2006). Conceptualizing rurality. In P. Cloke, T. Marsden, & P. Mooney (Eds.), *The handbook of rural studies* (pp. 18–28). Sage.

Depraz, S. (2008). *Géographie des espaces naturels protégés. Genèse, principes et enjeux territoriaux*. Armand Colin.

Gavinelli, D. & Marengo, M. (2021). Il gruppo AGeI di Geografia e Letteratura: Questioni di reciprocità dialogiche e territoriali tra produzioni letterarie e prospettive geografiche. *Geotema*, 66, 3–10.
Gigliucci, R. (2001). *Cesare Pavese*. Bruno Mondadori.
Hones, S. (2022). *Literary geography*. Routledge.
Jakob, M. (2009). *Il paesaggio*. Il Mulino.
Marchese, L. (2014). 'I Ciechi' di Cesare Pavese tra echi psicanalitici e sincretismo letterario. *Studi Novecenteschi*, *41*(87), 195–215.
Marengo, M. (2022). *Geografia e letteratura. Piccolo manuale d'uso*. Pàtron.
Meloni, P. (2023). *Nostalgia rurale. Antropologia visiva di un immaginario contemporaneo*. Meltemi.
Mencarini, B. (2013). *L'inconsolabile. Pavese, il mito e la memoria*. Edizioni dell'Orso.
Messina, G. (2022). Cesare Pavese's Il campo di granturco: Geographical trails. *Cultural Geographies*, *29*(3), 465–471.
Mondo, L. (2006). *Quell'antico ragazzo. Vita di Cesare Pavese*. Rizzoli.
Paül V., Tort, J., & Trillo, J. M. (2020), Presentation. Rural landscapes beyond the Idyll. *Documents d'Anàlisi Geogràfica*, *66*(2), 245–255.
Pavese, C. (2000). *Paesi tuoi*. Einaudi (original work published 1941).
Pavese, C. (2001a). *Lavorare stanca*. Einaudi (original work published 1936).
Pavese, C. (2001b). *Selected works of Cesare Pavese*. New York Review of Books.
Pavese, C. (2018). *The beautiful summer*. Penguin (original work published 1949).
Pavese, C. (2021a). *The house on the hill*. Penguin (original work published 1948).
Pavese, C. (2021b). *The moon and the bonfires*. Penguin (original work published 1950).
Pavese, C. (2021c). *L'opera poetica. Testi editi, inediti, traduzioni*. Mondadori.
Pettenati, G. (2019). *I paesaggi culturali Unesco in Italia*. Franco Angeli.
Roger, A. (1997). *Court traité du paysage*. Gallimard.
Turri E., Capozzi, B., Guadagnini, V., Negri, G. G., & Tadini, E. (2001). *Gli iconemi: Storia e memoria del paesaggio*. Electa.
Vaccaneo, F. (2020). *Cesare Pavese vita colline libri*. Priuli & Verlucca.
Woods, M. (2005). *Rural geography*. Sage.

Conclusion

Italian socionatures*

Filippo Menga

Introduction

In the last decade or so, the debate on socionatures has become quite vibrant in Italian academia. This book is a testament to that. In a way, I am also a testament to this shift – or rather, I believe I am well-placed to explain why I find the debate to be so vibrant. After working in the British academic system for several years, I returned to Italy in 2021. Until then, I was mostly familiar with the British/Anglophone debate on socionatures, having been shaped by my time in the Geography Departments at the University of Manchester and the University of Reading. Upon my return to Italy, one of the first challenges I faced was that it wasn't straightforward to find academic articles focusing on Italy that explained how nature in the *Belpaese* (see Papotti, in this volume) was essentially an artificial construct. This gap was particularly noticeable in the Anglophone knowledge production system, as the few works on the subject were mostly published in Italian. Now, there is nothing inherently wrong with publishing articles about socionatures in Italy in Italian – on the contrary, I think it makes a lot of sense. However, we can safely say that a debate only comes onto the international radar when it finally enters the Anglophone academic discourse. I am aware that there is an irony in the fact that journals labelling themselves "international" are almost always published exclusively in English. But however ironic it may be, the establishment of English as an international *lingua franca* in academia is a reality, and we have to come to terms with this.

Elsewhere (Menga, 2024), I explained how English clearly occupies a privileged position in the production of knowledge, including in the field of geography (for critiques on the use of English as a lingua franca at International Critical Geography Group conferences, see Belina, 2005; Bialasiewicz, 2003). We can think of this in terms of the "geographies of Geography" – both as a discipline and as a system of knowledge production. These geographies are diverse, fluid, processual, and marked by inequality, both globally and within specific regions or countries. As a political geographer focused on water, I see

* Part of what is included in the section "Why Italy, why now", and particularly the paragraph on populist ecologies, is a revised version of a section of an earlier article published in *Area* (Atkins & Menga, 2021).

DOI: 10.4324/9781003465164-23

clear parallels between the geographies of Geography and the geographies of freshwater resources: just as some countries have more water than others, certain regions within those countries are also more water-rich. Take, for example, the arid Northeast of water-abundant Brazil. Additionally, access to clean and safe water is often determined by factors such as income, class, race, gender, and age, affecting who benefits and who is excluded at any given time. While the impacts of water crises tend to be geographically localised and unequally distributed across communities and social groups, their root causes are structural. This kind of localisation and unequal distribution also applies to academic knowledge production and the varying pressures and challenges that geographers face – whether in publishing, accessing financial and intellectual resources, finding opportunities for mobility, or achieving professional recognition. We can thus observe that English dominance has two key elements: (i) the rise of English as the leading language in global knowledge creation; and (ii) the emergence of the Anglo–American world as the central hub for that knowledge.

The above may not seem directly related to the current book, but it is. Anglocentrism is in fact a two-way process. On the one hand, Italian academics tend to read and write (or at least have done so in the past) primarily in Italian, because this is the language they naturally use in national meetings. Those who work at the international level do the same, but in English. On the other hand, and precisely because of the former, Italian academics will deal with debates originating in Italian academia, and the international debate will be treated only marginally or based on outdated papers. The same will happen with Anglophone papers, and as a result, the Italian debate will be largely absent from Anglophone research. The only exception would be if an Italian academic working in an Anglophone institution or who has published his or her work in English adds references to Italian articles because he or she is more familiar with these debates. But that's really an exception.

So why is this book being published now, and why would it have been unlikely for such a book to have been published ten or fifteen years ago? The answer is twofold: first, Italian academia is evolving, and this book reflects that change. Italian scholars are increasingly proficient in both Italian and English, and while there is a common tendency to critique Italians' English-speaking and writing skills, significant progress has been made in this area. Second, the commodification of nature and the transformation of the Italian environmental landscape are more relevant than ever. This is the ideal time to explore these issues in depth. I will further explore this aspect in the following section.

Why Italy, why now

The environment, along with the redefinition of the very concept of nature, has never been as central to the Italian political and cultural debate as it is now. The 2021 rebranding of the Italian Ministry of the Environment to the "Ministry for Ecological Transition" (a post held by Roberto Cingolani), though short-lived – it was renamed the Ministry of the Environment and

Energy Security in 2022 – marked, at least symbolically, a peak in Italy's engagement with the diverse debates surrounding the climate crisis and global warming, initially sparked in 2018 by Swedish activist Greta Thunberg and the Fridays for Future movement in the Western world. Given that the necessity of transitioning away from fossil fuels has become widely acknowledged, the Italian government sought to place this issue at the heart of its environmental discourse, even though, in practice, little has changed. When Cingolani and Thunberg met at the 2021 Youth4Climate Summit in Milan, the Italian minister could not conceal his disdain for the Swedish activist. When asked about Greta's dismissal of global leaders' promises to tackle the climate emergency as mere "blah blah blah", he responded, "As long as [Greta's] protest remains a protest and not a proposal, it becomes part of the problem" (AGI, 2021). As an example of how the "ecological transition" is more achievable in theory than in practice, it is worth recollecting that in 2022, when Russia invaded Ukraine, Russian gas supplies vanished from the energy mix of many European countries, including Italy. In April 2022, ahead of an energy crisis, Italy's then-Prime Minister Mario Draghi posed a stark question to its citizens: "Do you want air conditioning or peace?". People chose air conditioning, and by winter, heating remained on as usual. As expected, energy prices soared, disproportionately affecting the poorest, but daily life continued largely unchanged. But the newfound energy security did not translate into an accelerated transition to renewable energy. One might have thought that cutting off Russian gas, oil and coal would have accelerated the EU's transition to renewables. Instead, it led to the strengthening of alternative gas import routes – mainly from North Africa – and the reopening of coal-fired power plants, even in Germany, the supposed environmental leader of Europe.

The need to study how the Italian capitalist society is appropriating and commodifying nature is made even more pressing if we consider the evolving political landscape of Europe, where the current resurgence of populist nationalism is as much an ecological issue as it is a political, social, and economic one. In recent years, geographers have indeed increasingly explored the connection between authoritarian populism and environmental governance, advancing our understanding of what can be termed "populist ecologies", broadly understood as the linkages between contemporary populism and nature–society relations (Atkins & Menga, 2021). As they examine how populism is reshaping humanity's relationship with nature, geographers question both the definitions of populism and nature itself, exploring how populist movements are influenced by, and in turn influence, the growing scientific, political, and popular debates around environmental degradation and sustainable futures in the Anthropocene. These studies reveal that populism transcends political and physical boundaries, expanding capitalist and extractive activities under a new guise. As D'Eramo (2013) observes, the decline of traditional political ideologies and party systems, combined with the normalisation of "apolitical" politics, where "the people" as a distinct political entity have largely disappeared from discourse, means that any critique of mainstream neoliberal capitalism is often branded as populist. Under these conditions, it is clear that environmentalism and any critique of capitalism – and consequently, of the capitalist logic of endless growth that underpins fossil fuel extraction and is behind uncontrolled environmental degradation – can also be quickly labelled as extremist or unreasonable.

Let us briefly consider how Matteo Salvini (the leader of the Italian party La Lega) tends to address environmental issues, something that he does in the context of supporting agriculture and protecting the interests of Italian farmers. For La Lega, environmental protection becomes an incidental – or even unintended – outcome of its nationalist agenda, reflecting a post-political populist ecology that discards traditional political ideologies, where the priorities of the "people" always come before those of nature. Unlike other right-wing movements like Fratelli d'Italia, La Lega's political platform does not prominently feature an eco-fascist link between people and land. This might be because the party originated as a secessionist movement centred on Padanian nationalism (La Lega Nord), rather than as an effort to resurrect the Roman Empire (la terza Roma). "Defending Italian agriculture", Salvini has stated,

> means defending the environment; defending local products means defending 'Made in Italy', with less diesel needed to transport goods and fewer imported products. Anyone still thinking about fascism or communism is stuck in the past! Let's think about the future, change Europe, and focus on jobs.
>
> (Lega – Salvini Premier, 2019)

In a similar vein, Vincenzo Sofo, a Member of the European Parliament from La Lega, argues that eating meat is beneficial for the environment because protecting the Italian meat industry – valued at €30 billion – also protects the environment (Lega – Salvini Premier, 2020). Here, the term "environment" (*ambiente*) does not refer to ecosystems or natural resources but rather serves as a populist catch-all for the Italian people who live, consume, do business, and vote within the country. In practical terms, this approach reinforces unsustainable environmental practices. The social and financial costs of environmental reforms, such as the European Green Deal, are seen as too burdensome for Italy's agricultural and industrial sectors, and environmental advocates are dismissed, with Greta derisively referred to as "Gretina" (from *cretina*, meaning "imbecile") and described as a "lunatic". The continuation of "business–as–usual" in terms of resource extraction and environmental degradation was also a key reason why La Lega withdrew its support for the Conte I Cabinet in the summer of 2019. At that time, Salvini clashed with the Five Star Movement (5SM) and Sergio Costa, the Minister of the Environment, over issues like offshore drilling and waste incinerators. As Salvini said,

> Everywhere else, when oil is found underground or under the sea, citizens and businesses celebrate because it means less dependence on other countries for energy. But in Italy, no – some people want to stop this, saying 'For God's sake, no, no, no'.
>
> (Rinnovabili.it, 2019)

Today, a few years (and one government) later, the debate on sustainability, the preservation of the Italian landscape, and the relationship between nature and society remains central to Italian politics – evident in the highly polarised discussions on wind energy in Sardinia (on Sardinia – but not on wind energy, refer to the chapters by Bonati and Mazza on animality in city branding, and Loi and Pierucci on urban green policies). This is the same political and cultural context in which the authors of this book are situated.

Social constructions of nature in Italy

And indeed, this edited volume explored the complex and multifaceted ways in which nature is commodified, constructed, and contested in Italy. Through a diverse range of case studies and theoretical approaches, the contributors examined how Italian natures were shaped, branded, and reimagined through policies, narratives, and practices. Although focused on the Italian context, the book provided insights relevant to broader global discussions about the relationship between nature and society in the Anthropocene era. Several key themes and patterns emerged across the chapters, and a central thread was the critical examination of the *Belpaese* (beautiful country) narrative and its role in constructing and commodifying Italian nature. This longstanding branding strategy, which emphasised Italy's natural beauty alongside its cultural heritage, had profound effects on how landscapes were managed, protected, and marketed. The book revealed that while this narrative celebrated Italy's natural assets, it often participated in neoliberal processes of nature exploitation and accumulation by conservation (Buscher & Fletcher, 2020).

The chapters also explored how natural spaces and features in Italy were transformed into "heritage" through various institutional and cultural processes. This heritagisation was examined in contexts such as national parks (Zanolin), UNESCO Biosphere Reserves (Cisani), food products (Pettenati), and urban green spaces (Loi & Pierucci). The authors demonstrated that these conservation efforts could lead to the commodification of nature and reinforce problematic nature–culture dichotomies. The volume also paid significant attention to the production and transformation of nature in urban contexts, highlighting how urban greening projects, climate adaptation strategies, and brownfield redevelopments (e.g., Garda) reshaped the relationship between cities and nature, often with uneven social and ecological consequences.

Many contributors have also addressed the conflicts and tensions that arise from competing visions of nature and landscape management, whereby case studies on the Alps, coastal areas, and agricultural lands reveal how different actors – from local communities to international tourists and investors – can contest the use and meaning of natural spaces. These conflicts often centre around issues of sustainability, climate change adaptation, and the balance between conservation and economic development. Rewilding in mountain and rural areas – a phenomenon that has been largely discussed in the anglophone literature – is also discussed in this book, with chapters (see in particular Piana,

Brocada, Dossche & Priarone), outlining how processes of land abandonment and the return of wildlife are creating new ecological and social dynamics. Rewilding, it is shown, poses both challenges and opportunities, including its potential to reshape human–nature relationships and contribute to nature-based solutions. The book also pays heed to literary and cultural representations of nature, paying heed to Geography's recent opening to the fields of ecocriticism and environmental humanities, and providing historical context for understanding how perceptions of nature have evolved in Italy, from the formation of national identity in the 19th century to the contemporary reimagining of rural and urban landscapes.

The spectre of climate change and wider environmental crises looms large throughout the volume – as it does outside the Italian setting – and the contributors examine how these global challenges are manifesting themselves in specific Italian contexts, from melting glaciers in the Dolomites to coastal erosion and changing agricultural practices. These works show how responses to these crises are shaping new understandings and management approaches to nature and raise questions about how we can think about alternative ways of conceptualising and relating to nature that go beyond traditional nature–society dichotomies. It is clear that a more nuanced, hybrid understanding of socionatures is needed to recognise the inseparability of human and non-human actors in shaping landscapes and ecologies.

Overall, this volume makes several important contributions to the study of nature–society relations, and provides a comprehensive and up-to-date examination of how nature is constructed, commodified, and contested in contemporary Italy, filling a gap in the Anglophone literature on Italian environmental studies. Its interdisciplinary approach – grounded in geography but incorporating insights from environmental studies, urban planning, and literary studies – offers a rich and multifaceted analysis of human–nature relations in Italy. And while focused on Italy, the book's themes and theoretical approaches contribute to broader debates in political ecology, urban studies, and environmental humanities, offering insights relevant to other national and global contexts. Thanks to this book, we now have a foundation for future research on socionatures in Italy and beyond.

References

AGI. (2021, November 21). Il ministro Cingolani ha parlato contro Greta e "l'ipocrisia del bla–bla–bla". https://www.agi.it/politica/news/2021–11–07/cingolani–contro–greta–ipocrisia–bla–bla–bla–14472258/

Atkins, E., & Menga, F. (2021). Populist ecologies. *Area*, *54*(2), 224–232. https://doi.org/10.1111/area.12763

Belina, B. (2005). Anglophones: If you want us to understand you, you will have to speak understandably. *Antipode*, *37*(5), 853. https://doi.org/10.1111/j.0066–4812.2005.00534.x

Bialasiewicz, L., (2003). The many wor(l)ds of difference and dissent. *Antipode*, *35*(1), 14–23. https://doi.org/10.1111/1467–8330.00299

Buscher, B., & Fletcher, R. (2020). *The conservation revolution: Radical ideas for saving nature beyond the Anthropocene*. Verso Books.
D'Eramo, M. (2013). Populism and the new oligarchy. *New Left Review*, *82*, 5–28.
Lega – Salvini Premier. (2019, April 8). #Salvini a Quartarepubblica: Difendere l'agricoltura italiana significa difendere l'ambiente. Facebook. https://www.facebook.com/legasalvinipremier/photos/a.423252457717686/2301492069893706/?type=3andtheater
Lega – Salvini Premier. (2020, July 7). M5S contro i produttori italiani: In commissione a Bruxelles europarlamentare propone tassa sulla carne per "tutelare l'ambiente". Follia! Facebook. https://www.facebook.com/legasalvinipremier/videos/276186403487022
Menga, F. (2024). Multiple political geographies. In F. Menga, C. Nagel, K. Grove, & K. Peters (Eds.), *Political geography in practice: Theories, approaches, methodologies* (pp. 225–238). Palgrave.
Rinnovabili.it. (2019, July 31). Salvini: "Si a termovalorizzatori, TAV e trivellazioni in mareoaterra". Rinnovabili.it., https://www.rinnovabili.it/ambiente/salvini–termovalorizzatori–trivellazioni/

Index

Pages in *italics* refer to figures, pages in **bold** refer to tables, and pages followed by "n" refer to notes.

Abruzzo 19, 44, 46n7; national park 15, 17, 19, 22
accumulation: of capital 118; by conservation 7, 27, 61, 82, 227; through conservation 34; mechanisms 110; by sustainability 91–92
agrivoltaics 108–116, 119
agro-silvo-pastoral 37–38, 42
Alps 8, 42, 44, 73, 83–84, 89, 152, 227; Central-Western 92; eastern 73; Eastern Italian 82; Italian 49, 56, 91; Julian 29; Ligurian 8, 38, 42, 44; Orobic 55–56; pre-Alps 129; Prealps 154, 156
Ampezzo valley 83–84, 87, 91
authenticity 8, 53, 61–63, 65–69, 178

beautification 4–6, 10, 95–97, 104, 163, 168
Bel Paese 4, 7, 9–10, 15, 95, 109, 119, 191–193, 195–196, 198n11, 198n15–198n16, 198n18
Bergamo 127, 129, **130**, 132–134, 136, 154, 157–158, 160n3; Hills Regional Park 154
Bitto 55–56
botulinisation 5–6
branding: city 9, 174–178, 182, 184, 227; food 51; mountains 91; narratives 7; nation 212; place 5–6, 51, 176; practices 5, 176; processes 95; rebranding 127–128, 133–134, 224; theory 6
Brescia 127, 129, **130**, 133–134, 136, 157–158; Hills Park 154; Quarries Park 159–160
brownfield 9, 139, 151, 227

Cagliari 9, 163–164, 166
Campania 96, 105
capitalisation 3; nature 7; of nature 6
Central-Western Alps 92
climate: adaptation 227; adaptation and mitigation 133; change 8–9, 15, 32, 73, 76–77, 79, 85, 89–90, 96, 99–100, 103, 108–110, 118–119, 125–127, 129, **130**, 132–134, 140, 146–147, 228; change adaptation 158, 227; change adaptation/mitigation 7; change resilience 114; mitigation 108; resilience 133; transitions 129, 132–133
coastalisation 9, 95–96, 100, 104
commodification: of agricultural landscapes 108; and capitalisation of nature 6; of Italy 3; of mountain nature 84; of mountains 83; of natural resources 119; of nature 9, 23, 51, 89, 96, 110, 112, 118, 224, 227
conviviality 34, 152, 156
Cortina D'Ampezzo 84
cultural: commodification 176; heritage 5, 29, 32, 34, 57, 64, 68, 90, 227; identity 66, 167, 182, 191; landscape 32–33, 38, 45, 157, 174, 214
Cungiaus 9, 164, 166, 171, 172n2

desealing 9, 139–140, 144, 146, 148
Dolomites 8, 73, 75, 77, 82–85, 89–92

Eastern Italian Alps 82
eco-climate crisis 82–86, 90–92

ecology: glacial 78; political 3, 15, 82–83, 91–92, 125, 228; populist 226; urban political 9, 125–126, 152, 164, 168
exploitation: of the glacier 74; of protected nature 34; of nature 82

fiorentina 8, 62–64, 67–68
Florence 61–68
food: biopolitics 57; branding 51; culture 52–53, 55, 57; foodification 64, 68; foodscape 51, 57; geography 49; heritagisation 49; mountainness 49

geo-literary 9, 212
glacial landscape 73, 77
grabbing 9, 108–110, 112, 115, 119
green: beautification 6; extractivism 118; gentrification 158; grabbing 9, 108–110, 112, 115, 119; greenwashing 118; island 96, 101–102; mobility 91; nature capitalization 82; tourism 45, 89

heritage: conservation 152; exploitation 8; Italian literary 192; making 27–29, 34, 61, 63–64, 66, 68–69; reproducing 7

ice landscape 76–77, 79
Ischia 96, 101–104
Italian Alps 49, 56, 91
Italian landscape 5, 7, 9, 191, 212, 227

landscape: commodification 38; conservation 118, 158; grabbing 110, 119; -as-heritage 27–28; management 227; transformation 95, 175; urban 165, 228
Langhe 214–216, 218, 220
Lazio 15, 32, 169
Ligurian: Alps 8, 38, 42, 44; Apennines 42–43; landscapes 37–38
Lombardy 43, 55, 129, **130–131**, 143, 151–154, *155*, 156, 158–160, 192

I Malavoglia 200–204, 206–207, 210n2, 210n5
Mantova (Mantua) 127, 129, **131**, 132–134
Marche 96–98, 105
Marmolada 8, 73–80, 84, 86
Mediterranean Sea 10, 43
Milan 116, 141, 143, 145, 147, 153, 160n3, 163, 192, 196, 225
Milano Cortina 2026 82–84, 86–91
Molise 15, 32
more-than-human: agents 57; contemporary thought 2; geography 2; literature 174; perspective 184; space 174; transformation 50
mountain: commodification 90–91, 96; food culture 57; geography 83; landscape 45; Ligurian 46; mountaineering 75; nature 83, 88–92; parks 153; pastures 54–56; system of Monte Antola 42; symbolic 73; -urban 83
Mountainness 49, 57–58

national: identity 10, 15, 17–19, 23, 191, 193–194, 196, 198, 228; landscape 10, 197; nationalism 15, 225–226
natural landscape 6–7, 20
naturalness 8, 49–52, 56, 62, 67–68, 151, 158, 165, 194; cheese 49, 52
nature: based solution 6, 9, 128, 140, 228; commodification of 9, 23, 51, 89, 96; exploitation 7, 90, 171, 227; peri-urban 65, 160; rewilding 37; social 1, 16, 27, 49, 109, 119, 152, 174; urban 125–128, 132, 134, 136, 158, 163–169, 171; wild 15, 23, 151, 163–165, 169

Oristano 174–181, *184*, 184n1
Orobic Alps 55–56

park: local 153, 155; national 8, 15–21, 23, 29, 32, 151, 227; natural 38, 84, 155; park-making 163–166, 168–171; peri-urban 159; photovoltaic 156, 158; regional 29, 31, 153, 155, 158; riverside 168–169; suburban 153; urban 9, 140, 142–143, 145, 151
Pavese, Cesare 10, 212, 215–216, 218, 221, 221n2
performativity 65, 69, 174–178, 180, 182, 184
planning tools 142, 146, 160
protected areas 8–9, 15, 26–29, 34–35, 43, 51, 151–157, 160

Quartucciu 9, 164, 166–167, 171, 172n2

renaturalisation 6, 9, 50, **130**, 132, 139–141, 147, 159, 164, 166–168
renewable energy 108–110, 117–119, 119n1, 133, 156, 158, 225

rewilding 37–38, 43–46, 46n1, 46n7, 151, 227–228
Romanticism 18, 151, 164
Rome 5, 9, 22, 116, 163–164, 166, 168–171, 191
rural: idyll 212–214, 221; landscape 10, 32, 52, 67, 132, 155–157, 159, 212–214, 218, 220

Sardinia 174–177, 179–180, 184n1
Sartiglia 174–182, *183–184*, 184n1
Senigallia 96–100, *99–100*, 104
slow food 49, 53, 55
socionature 1, 58, 125, 127–128, 132–134, *135*, 136, 152, 155, 158–160, 223, 228; urban 125, *135*, 136
Stoppani, Antonio 10, 191–196, 198n12, 198n18

Tevere: Master Plan 168, 170; Parco Marconi 168–169
third landscape 145, 151
Tiber 9, 164, 168–171
touristification 63–64, 95; hyper- 104; urban 61
Trentino 76; Alto Adige 84, 89
Trento 141, 146–147

UNESCO: biosphere reserve 227; MAB 30, 33; World Heritage List 214
urban: climate strategies 127, 129, 136; geography 9, 134, 166; imaginaries 96, 174; metabolism 125, 129, 141, 169; mountain- 83; nature 125–128, 132, 134, 136, 158, 163–169, 171; park 9, 140, 142–143, 145, 151; peri- 38, 65, **130**, 132, 151–152, 154, 156, 158–160; political ecology 9, 125–126, 152, 164, 168; sustainability 125–126, 128, 163; void 139, 143, 145, 147–148, 164
urbanisation: coastal 96, 159; extended 126; hyper- 95, 104; nature 104; of nature 127, 129, 152, 159–160; planetary 126–127; sand 96, 104; tourism 95

Veneto 76, 84–89
Verga, Giovanni 200–209

wild: encounters 163, 171; landscape 9, 102; nature 15, 23, 151, 163–165, 169
wilderness 2, 17, 20, 32, 43–45, 46n2, 75, 164, 219
wildlife 2, 8, 39, 44–45, 228
wildness 164–169; urban 151

Kolibri 360 GmbH
Ferdinand-Jühlke-Straße 7
99095 Erfurt, Deutschland
produktsicherheit@kolibri360.de